Some Airplane Stories and History of a WWII Pilot in the Aleutians

Pat Patteson

Our skipper, Lt. Cmdr. Paul Staul, USN, was a graduate of the U.S. Naval Academy. While I don't think he was in the top percentage of his graduating class, by all other standards he was tops. He flew missions with us until grounded by the Commodore, Gehres. He had our respect — both the officers and enlisted crew. And he respected we the 'Reservists' in return. He was proud of his squadron and wanted to show this by ordering a special patch made for his squadron personnel.

This patch is a small emblem that airmen wear on the breast of their flight jackets to identify their unit. It shows who we are and that we are active in combat flight duties. It can get you attention and respect. Others recognize it. My good friend Tom Macleod, another pilot volunteered to do a representa-tion. This is what he came up with-.

A blind blue fox astride a flying gasoline tank carrying a bomb and several machine guns through the fog over the volcanic Aleutian Islands. The opposite paw holds a cane to assist in navigating through the clouds.

Thumbs Up, from our skipper.

Some Airplane Stories and History of a WWII Pilot in the Aleutians

Pat Patteson

Edited by Don & Denise Downey

WWII Veterans History Fund
2022

First Printing: 2022

ISBN 978-1-63-010030-8

Library of Congress Control Number: 2022920755

WWII Veterans History Fund
San Ramon, California 94583-3004

www.dondowney.com

Dedication

To my son, Arthur Lea, for his love and devotion that has added so much to my life.
We happily support each other mentally and physically in warm comfort.
I am thankful to hundreds of my true friends that have entered my life's journey.
Too many to try to name.

But, my family is first as they formed the core ethics of
Love, Devotion, Work, Play, and Purpose.
We four children were all raised in a small farm and ranch community
when two world-shaping events occurred; the Great Depression and World War II.

By our parents love and sacrifices they managed to send all of us to Colleges.
This is to chronicle part of my experiences in the US Navy in WWII.

Lea and Pat at the 3rd Thursday Denny's Veteran Lunch Bunch in Concord, CA.

Table of Contents

Acknowledgements

I'm proud of a long list of friends who deserve my utmost thanks. I will list just a few —but to list all would be difficult.

First, my son, Lea. He has always wanted to know more of my early life at Smiley, and he liked to visit and talk with his grandpa. Then, he is curious about my war experiences in WWII. I hope this book pleases him.

To Taigh Ramey, who let me fly that old warbird, a PV-2 Lockheed 'Harpoon.' It revived my memories so vividly — to hear that throaty roar of those powerful Pratt & Whitney R-2800 engines and the intoxicating smell of 100-octane fuel. My memories, whether good or otherwise, rushed back to mind instantly. I felt like a pilot again.

To Dr. and Mrs. Boris Inchenko and their talented daughters, who opened their doors of hospitality to Lea and me. He researched and wrote articles in air magazines to describe what was going on in the "Forgotten War" when all eyes were on Europe. I have enjoyed years of friendship with these friends.

To Gil Ferrey, a veteran helicopter pilot of the Viet Nam war. Gil has given me a revealing review of the activity and effect of that patriotic effort to stop Communism. Gil is known as an encyclopedia of aviation. I appreciate his help and concern for Lea and me to keep us up to date with many vet activities in the local area. We attend and enjoy all.

And the Methodist Church group called 'Westley Writers' who meet and help each other. Most of us are struggling with the written word and how to use a comma. I certainly need help. As an amateur, it has been fun and eventful.

Two great aviation friends, Boris Inchenko and Taigh Ramey meet in Orlando, Florida.

Distinguished Flying Crosses (2)
Air Medals (2)
Asiatic–Pacific Campaign Medal
(Gold Stars for 2nd Medals)

Foreword: U.S. Navy Squadron VPB-135 History

During World War II the remote Aleutian Islands, home to the Unangax̂ (Aleut) people for over 8,000 years, became a fiercely contested battleground in the Pacific. This thousand-mile-long archipelago saw invasion by Japanese forces, the occupation of two islands; a mass relocation of Unangax̂ civilians; a 15-month air war; and one of the deadliest battles in the Pacific Theater. [National Parks Service website]

Pat Patteson was in it. *"We felt as if we were knocking on their (the Japanese) back door. I think we were more a threat than they thought we were."* Pat & Lea have a collection of Pat's WWII history including complete squadron VPB-135 records and dozens of never seen before photographs. Pat and Lea have presented the highlights in this book. [Don Downey, editor]

Recorded 19NOV2021 at his 100th birthday party, Pat, along with several other WWII veterans including Lou Gibbs (pilot), Jake Larson (Normandy) and Diz Laird (Navy Ace), passed around the microphone and gave a quick summary of their WWII service.

Don: Okay Pat, thanks for having us here and Happy Birthday Go right ahead and just say anything you'd like everybody. Thank you.

Pat: I'm Pat Patteson. I'm awaiting my 100th birthday. I was a Navy pilot and graduated with my gold wings from Corpus Christi, Texas, and was immediately sent to the Pacific and ended up in the North Pacific after training at Whidbey Island in a new airplane to the Navy. This time, instead of PBYs, they were going to introduce a land-based patrol plane, the Ventura, and we were formed into a 15 plane squadron and replaced most of the PBYs that had been on patrol in the Aleutian Islands. We were sent to that first, to the Aleutian Islands. It was Adak, then Amchitka, then Attu. And we flew photo reconnaissance. That was our job.

Our intelligence people at the Pentagon had very little information about the Japanese forces in the Kuril Islands. They had just taken the islands of Kiska and Attu. This was six months after Pearl Harbor. I flew in three photo reconnaissance squadrons, all of them VPB-135. In 1943, I came back, reformed with new pilots and new planes; in 1944, better equipped with more photographic equipment and, finally, in 1945, in a PV-2 squadron formed again and trained at Whidbey Island.

When the war was nearing its end on VJ Day, I was on Attu and most of us stayed inside our huts rather than get out where the celebrations were going on. I was one of the first to return and be separated from the Navy. I flew a plane all the way from -- with a copilot and a crew of other seamen and officers, back to the states. They had been eligible for discharge and I took them and dropped them off from Seattle to Burbank to Phoenix to El Paso and then into Abilene, Texas. So my war days were settled very quickly and I returned to school. I had been taken out of college after two years of credits and I got to civilian life and after my sojourn with the Navy. Thank you.

Introduction: Patteson Family Photos

Here are some of Pat's family photographs. The following sections include Pat's short stories, his collection of WWII photographs including many that he took himself, and transcripts of several interviews recorded in 2021 and 2022.

Photo 1: 11-24-1921 Artesian water from well for Dad's cattle herd - Smiley ranch. Dad and I inspect his water well.

Photo 2: A.L. “Buff” Patteson 141st Infantry Company ‘L’. Landed Brest, France, WWI 1917.

Photo 3: Our Mom in a rare vanity pose - managed to get her 4 kids college degrees. Mary Kercheville came to Smiley as a grade school teacher. Dad married her when he returned from France, a 2nd Lieutenant, WWI Infantryman.

Photo 4: Nelda Jean and Mary Beth - my younger delightful sisters pose for Smiley High School Album. Both were a happy delight and fun to be with. I missed their most charming years.

Photo 5: Jane Kercheville, Mary Beth, and Nelda students at Abilene Christian College. Jane, my cousin, eventually became an A.C.C. professor.

Photo 6: Brother Morris in premed attire. A top student, worked hard and earned his M.D. degree. Highly respected doctor in Lampasas region. Due to WWII we rarely commuted or had times to commute.

Photo 7: Morris and Dad “Buff”, at home. Note the Stetson hat.

Photo 8: As a Navy Cadet, I came home some weekends. Attended the Smiley Methodist Church with Nelda on those occasions.

Photo 9: The Twins at home 1942. Our Smiley visits were rare. I was a Navy cadet. Morris, a "med" student had our preoccupations.

Photo 10: Morris, my twin brother at Galveston Bay during final year at Texas Medical School. He was a hellova good looking guy!

Photo 11: This is not supposed to be funny. Morris was issued the clothing when he signed up for the Navy Medical Corps.

Photo 12: 1942 - Family weekend visit to Smiley. Finally, a commisioned officer.

Photo 13: Dad with his generation of sisters and brothers. Dad's oldest brother was named Walter. He or his family are not shown.

Photo 14: Post-war family.

Photo 15: Post-war photo Jolea and Pat. We loved and enjoyed 57 years of marriage. Never has there been a disease, as cruel, as sinister, as merciless, as Alzheimer's disease. Never! What it does to those victim's families, their loved ones, their caregivers, it is horrible beyond words to describe.

Photo 16: Jolea at Hawaii, our first squadron reunion.

Photo 17: Jolea made this oil painting of our view of Mt. Diablo that still hangs at home.

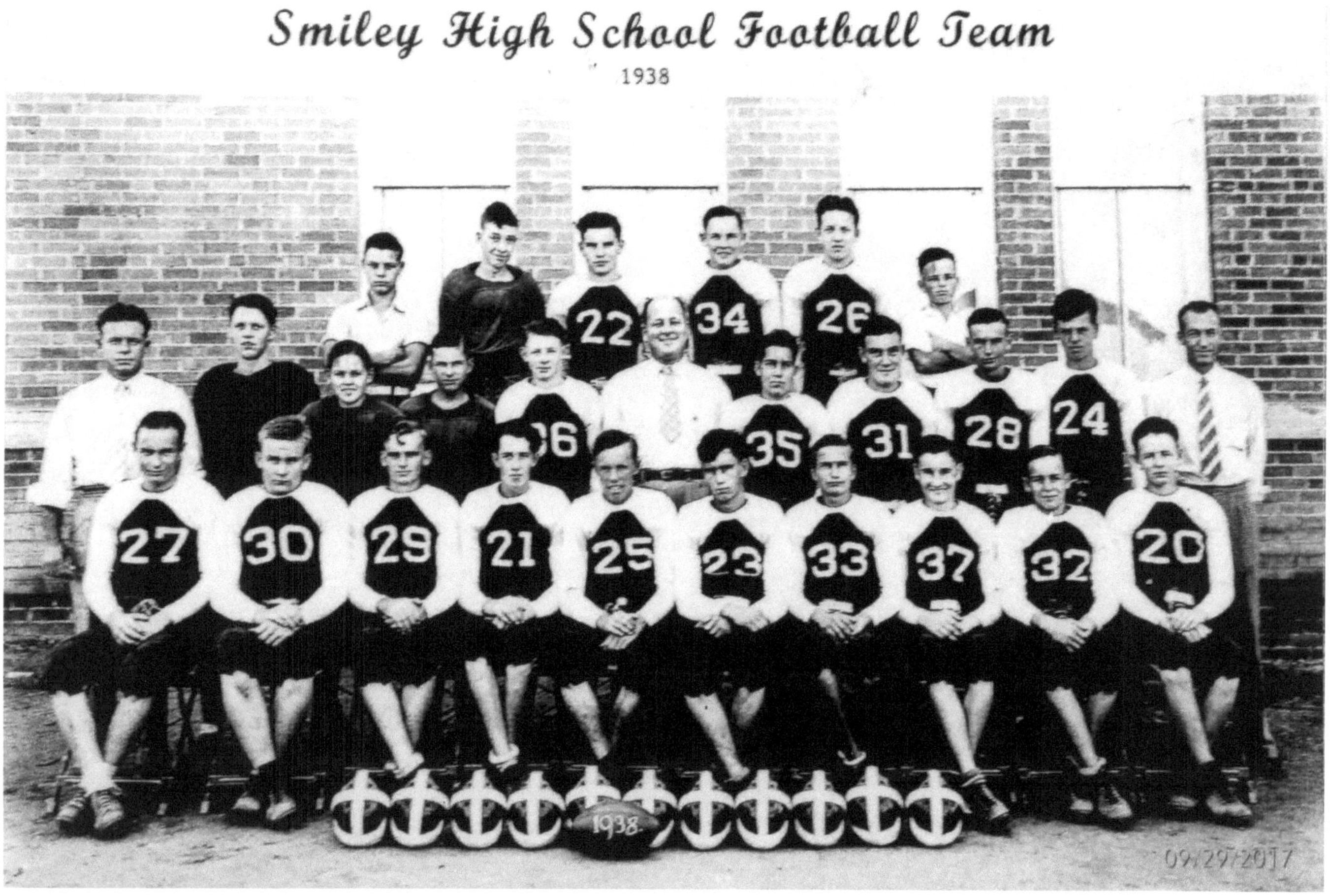

Photo 18: The mighty Smiley 'Bantams'. We won our division!

Preface: Maps of the Bombing and Photo Reconnaissance

There is so much to cover for a good story (or stories). I have an excellent report written by an intelligence officer of one of our combination photo runs and bombing missions that covers several pages of facts and observances. I want you to read it next visit.

Our missions were primarily for reconnaissance, the bombs were not dropped by use of the bombsight, and we rarely flew in formation. Each plane navigated on its own and was assigned separate targets to photograph. Strategic bombing was the job of the 11th AF.

We often carried a variety of bombs (general purpose, fragmentation, incendiary) and used them when we had good targets of opportunity. The 11th Air Force with their B-25s and B-24s was supposed to attack the land bases using the Norden bombsight. Weather often limited their opportunity and some of their big bombing raids were futile--the bombs were simply salvoed . Also, due to the long distance to the targets fighter cover was impossible.

Weather was rarely in our favor for photos or bomb runs. However, it also affected enemy fighters as their pilots had to stay on the ground.

Photo 19: VPB-135 crew at Naval Air Facility Attu (Casco Cove) ready to board their twin-engine PV-1 Ventura medium patrol bomber for another mission against Japanese airfields and ports on Paramushiro.

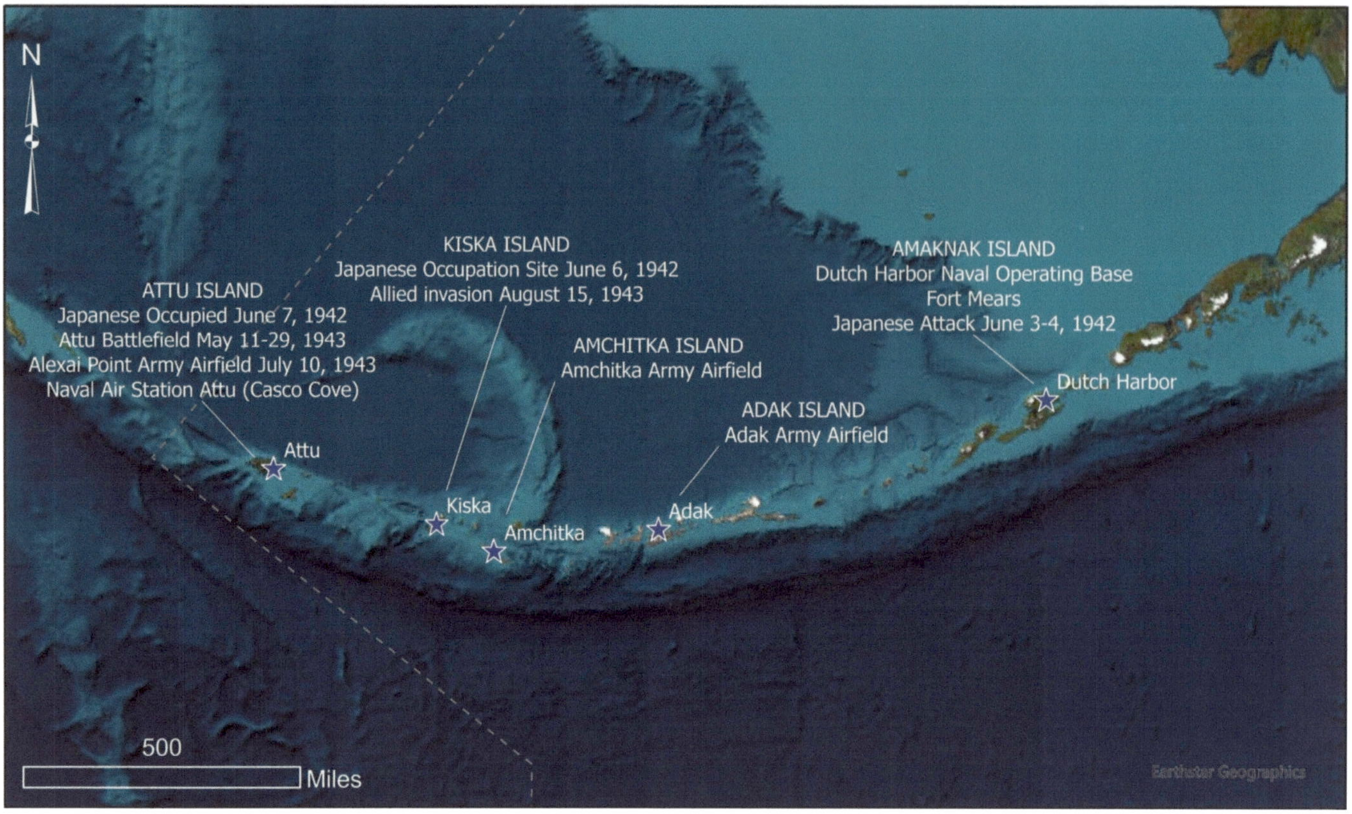

Photo 20: Aleutian Islands WWII Index Map (based on National Park Service map). Pat's VP-135 squadron flew bombing and photo reconnaissance missions out of Naval Air Station Attu.

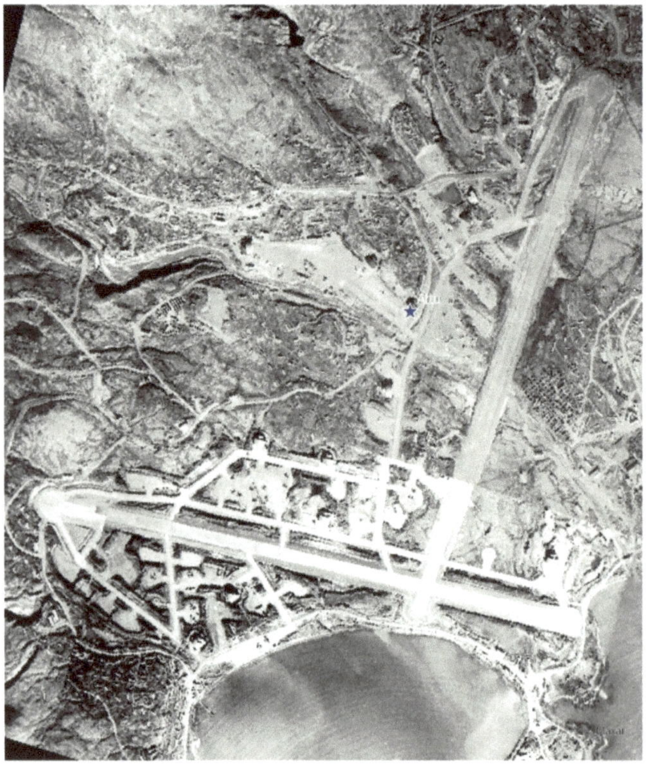

Photos 21a, 21b: Pat's Attu airfield 1944 and present day. Established as Naval Air Facility Attu on June 7, 1943, just seven days after Attu Island was declared secured. Later used as Casco Cove Airport LORAN naviga-tion station (United States Coast Guard). The island is protected as part of the National Parks Trust territory and few people visit. Gun emplacements for over twenty guns are still present around the airfield.

Photos 22a, b, c: Naval Air Station Attu (WWII) and former Coast Guard Station Casco Cove (present-day).

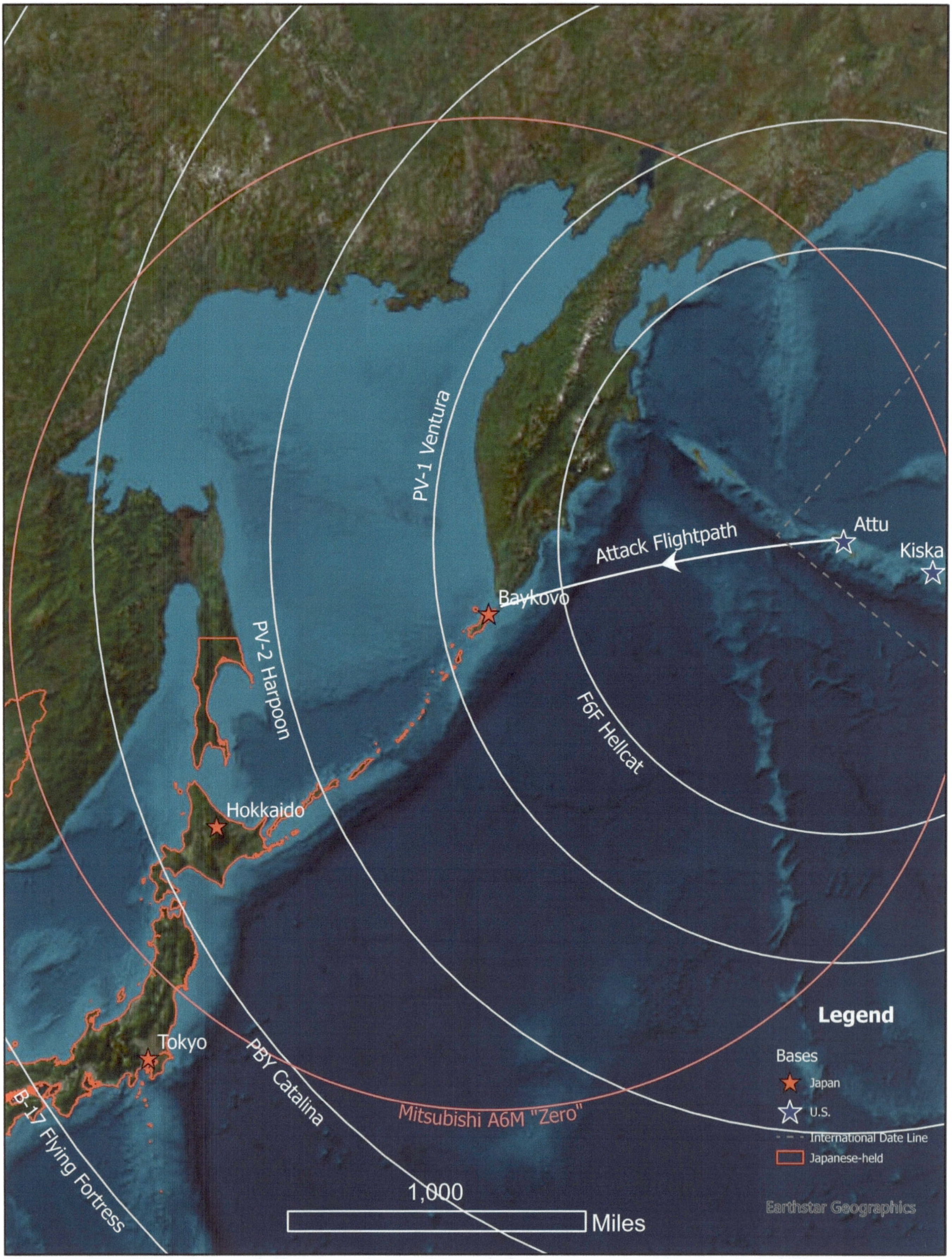

Photo 23: The PV-1 Ventura had a range that could reach from Attu to the Japanese-held islands of Shumushu and Paramushiro.

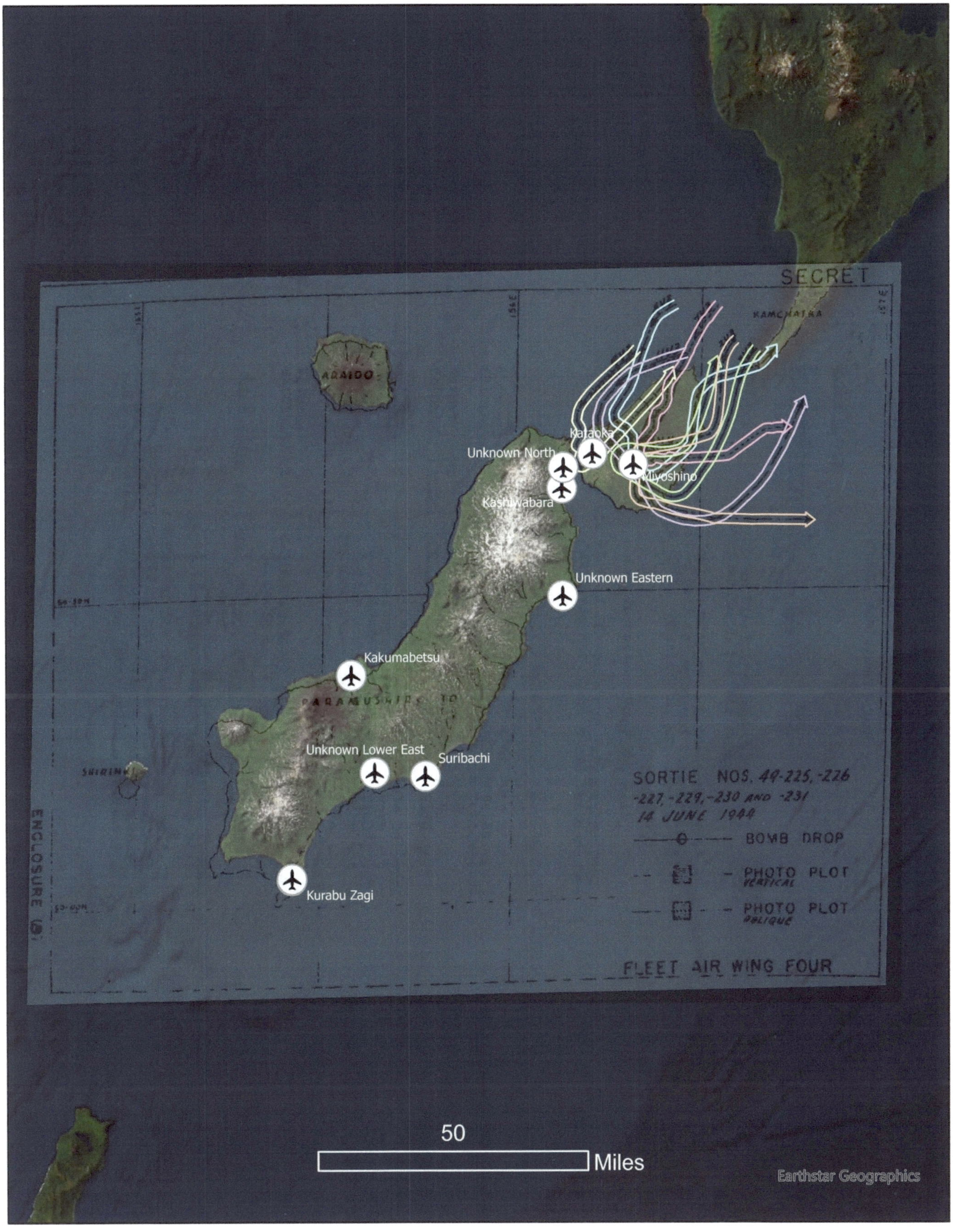

Photo 24: Sorties 49-225, 226. 227. 229. 230. 231 14JUNE1944 against Japanese bases and air fields on Shumushu Island.

Photo 25: Sorties 49-228 14JUNE1944 against Japanese bases and air fields on Paramushiro Island. The mission returned aerial photographs over the areas indicated in blue along the flightpath.

Broiler Capital of Texas

Photo 26: Recent photo of sign in Smiley, Texas.

It was hand-to-mouth in the dark days of the Big Depression. But, like most small farm and ranch communities they were near self-sustainable. Nearly every household had a water well, a garden patch, fruit trees, a cow, and chickens. A sewing machine and butter churn were household items just like the fireplace. I can't recall anyone going hungry. They just didn't have any money. Most families still had dreams of sending their kids to college. My Mom wanted me to become a violinist and my twin brother to be a doctor. Fortunately, Morris did become a prominent doctor and in a sense, I also found my dream, but not with the violin.

Beef was the money crop and revenues from those raising cattle kept the local bank alive. The village survived because of the cattle market and supported two local grocers, three filling stations, a restaurant, churches, and a garage for local car repairs. Most clothing and other household needs were bought from mail order and the Sears Roebuck catalog. At about that time a popular restaurant chain started that specialized in country fried chicken. Its success depended on serving fresh, tender, broiler meat, chicken fried with their secret (and patented) formula.

A Smiley man named Bell contracted to supply live chickens to the restaurant overnight by truck. This worked for a while until the fried chicken demand expanded and Mr. Bell had to get more help. To meet the restaurant's demand Smiley had to hatch more eggs, raise more broilers and build more chicken houses. A local industry sprung up.

At home, Barney and Aunt Snooks lived just across the fence from us. Our kin was anxious to get in on the chicken business and he had enough land to build seven chicken houses. Each house could hold a thousand chicks. Barney proposed a deal with me and my brother on a 50-50 basis until the broilers went to market, he would provide all the financing and we would do all the work. We cleaned, watered, fed, and accounted for the dead, sick, and disabled critters from baby chicks to broilers every day and night of their lives. Until it was D-Day eleven weeks later. The price of broiler meat fluctuated so selling on the wrong day was a gamble. Morris and I had dreams of bicycles, guns, and skates until one day there came a disease called coccidiosis. It spread from farm to farm. In a matter of days this virus could infect birds, houses, and even the soil. Farms that couldn't afford the chemicals to stop this epidemic were wiped out.

It became necessary to sell the healthy birds as soon as possible and this caused a glut on the market--the prices plunged. When our brood was ready for market the price was low. Barney lost his shirt like most others. Somehow to console us for nearly 12 weeks of work he paid us $10 each.

(Hence the term "chicken feed")!

Butterflies in the Arctic

Photo 27: PV-1 Ventura waits in winter snow at Attu base.

This is our first squadron mission--we were to get night photographs of the Japanese installations on Paramushiro. Ens. Walsh, Ens. Rice and I were roused from our sacks at 2000 by a marine who entered our hut carrying a clipboard and Garand rifle. He flashed a light in our face and thrust the clipboard at us, "Sign here, sir". Then left. Since we had already laid out our heavy flight gear before trying to get some sleep it took only a few minutes to dress. The rest of our crew; a turret gunner, radioman, and plane mechanic, would join us at the plane. Two miles away we could hear the acceleration and throaty roar of the big Pratt & Whitney engines being run up. Each engine could develop 2000 HP and was as much power as our latest fighters in the Pacific.

A small, neat, and tiered amphitheater had been built that could hold about fifty people. The Commodore and five of his aides entered the briefing room promptly at 2200 and took front seats. The briefing began with a friendly welcome by the PatWing Four Operations Officer who stressed the urgent importance of our mission. Seems Chester Nimitz (CincPac) needed intelligence vital to the invasion plans which included the North Pacific. To date, the high brass knew practically nothing about this Japanese stronghold. Our next briefer was the Wing's weatherman with his charts. He presented a lot of curvy lines (isobars) that extended into northern Siberia as well as our flight route. He noted that much of his synopsis was made up of projected weather data since we never received any weather data from our ally, Russia. This was followed by our plane assignments and armament load. Each plane was to carry two 500-lb GP bombs in the bomb bay and 12 magnesium flash bombs in the aft cabin. The camera, a Fairchild K19, was mounted in the nose to take vertical pictures. The flash bombs were manually set to go off by a timer so many seconds delay according to the altitude. Each flash bomb was dropped by hand one by one. The one million candlepower flash triggered an electric eye of the camera. Each bomb weighed 30 pounds; one man cradled the bomb while the other man set the timer and pulled the safety pin. A Navy Ordnance Chief had instructed the crewmen how to do this without fumbling.

As the briefing continued we received call signs, emergency frequencies, IFF settings, and a small vial of 'No Doze' pills. These were entrusted with the pilots to be issued en route, if necessary. We had a case when a crew was given the little pills before they took off only to find their plane had mechanical trouble. That whole crew remained wide awake for the remainder of the night and most of the next day. After a few briefing questions, we were given current code books in a weighted satchel, a thermos of hot coffee, and a bag of sandwiches. An old yellow school bus idled outside the briefing room for travel to the flight line. The senior pilots were dropped off to fly their normally assigned plane and my crew (#18 the most junior crew) took PV-1 #48933 usually flown by Bill Clapham, a senior pilot. Our call sign, 8V, proved to be a most reliable and maintenance-free plane. Bill and I would use it with confidence on many more missions. Finally, the briefing was over, the Commodore stood up, gave the two-finger V-sign, wished us 'good hunting', and left the room.

Photo 28: Squadron VPB-135 crew ready to board their PV-1 patrol bomber for another mission to Paramushiro. Plane is parked on Marsden Matting at Alexai Point Army Airfield , Attu, Territory of Alaska. Naval Air Facility Attu (Casco Cove)

When we arrived at the flight line our crew was already in their parachute harness and the radioman was doing his checkout. My AMM went with me as we did a routine walkaround and then we climbed aboard, buckled into our seats, and closed the hatch. Both engines, still warm from their tests, started without a cough or sneeze and we were ready to taxi. It was nearly a mile to the runway and the overloaded plane needed lots of throttle jockeying to stay on the rough taxiway. It was a dark night and nearly a mile to the runway. A fuel truck was waiting to top off our main tanks, every gallon counts and we had used over 40 gallons just getting started. The runway lights had been dimmed and our six-plane mission would take off about 1-1/2 minutes separation. Except for the pilot infrareds, all other lights were off. Our final checkoff list for takeoff had to be by memory and touch. The engines were gunned, brakes released, full right rudder applied against prop torque, and rotation came near the end of the runway. We were on instruments at 300 feet altitude and began a slow climb. Our course to the target, WSW.

It will take over 4 hours to reach our target. The trip is dull, dark, cold, and noisy. No one relaxed. Pilots scanned the instrument panel and occasionally tweaked the autopilot to adjust for any gyro precessing, the radioman stayed bent over his radar scope and listened to static on the watch frequencies, the navigator reviewed his calculations and plots. The gunners were trying to bide time and stay warm. Heavy gloves were necessary, don't touch anything metallic. I suspect some may have breathed a 'Hail, Mary', or two.

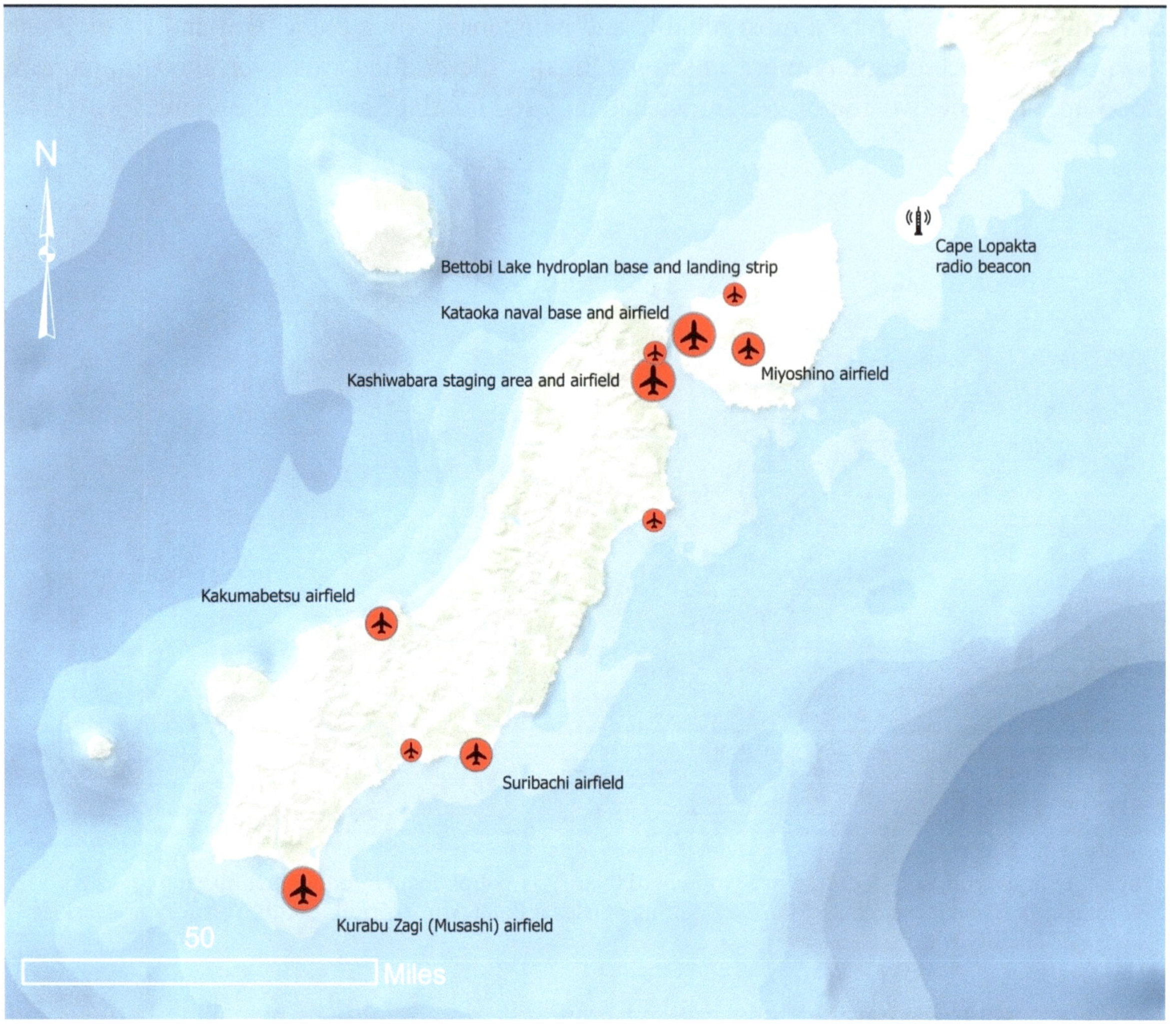

Photo 29: Cape Lopakta radio navigation beacon is located on the southernmost point of Kamchatka Peninsula, Russia and lies about 7 miles north of Shumushu, the northernmost island of the Kuril Islands.

Eventually, we picked up the faint 'LK' signal of Cape Lopatka, the southern tip of the Russian Kamchatka Peninsula. Four miles south of the beacon was enemy territory. As we go closer we saw cloud layers along the coastlines were covering the first Japanese island of Shumushu. Turning south WX had broken up over our target area and we could see flashes from AA ground batteries. Closer, and we saw the AA bursts were colored red, orange, and blue when they exploded near another of our planes. Meanwhile, we began dropping the flash bombs and suddenly the searchlights caught us and apparently also triggered the cam-era. Most AA bursts were behind us and caused no damage but nothing could shake those blinding lights, for miles they followed every move we tried. It took several minutes to dodge and get out of the area. Let's get the Hell out--head back to Attu satisfied that we had done our job.

About eight months before our squadron arrived on Attu the 11th Air Force made a surprise attack on the North Kuriles with some success. It surprised the Japs and the B-24s and B-25s had a successful bombing raid. Results could not be observed at the time due to cloud cover. A few 'Oscar' fighters managed to intercept and chase after the drop but did little damage to the bombers. The AA was light. Later, encouraged by this good raid, a bigger attack was planned for 6 B-24Ds and 12 B-25C/Ds. The heavies were to drop from high altitude and the B-25s would strike low at shipping and other surface targets. This time the surprise was different. Fresh, seasoned fighter pilots in Zero 'A6Ms' had been brought up from the Central Pacific plus more AA emplacements and even some night fighters. These veteran fighters used the tactic of head-on two-plane attacks against the slower bombers--hit, climb, circle, and hit again. They knew the B-24 is lightly gunned or armored in the nose so they had a field day. The Mitchell bombers making deck-level runs on cargo and fishing boats found there was heavier AA and small arms fire everywhere. It was brutal. The enemy won by a landslide--we lost 3 B-24s and 7 B-25s. And some of those that made it back to Attu base had wounded aboard. No more daylight strikes for this 11th Air Force until the Summer of 1944. They had had it...morale was the pits.

As we were leaving the coast of Paramushiro we sent a coded radio report to confirm we had photographed and bombed the target. We were now headed for home and had survived our first baptism of fire with hardly a scratch. The WX was blotchy but in about two hours we should be able to get a navigational bearing giving a good accurate heading home. Our next sweat would be fuel and base WX at Attu. The jinking about had used more fuel than we expected and reduced our margin of finding an alternate field. The nearest alternate landing strip to the main base was usually affected by the same WX as the base. As we came to the 100-mile mark a recognition circle was flown so our IFF would give a positive identity that we were friendly. Just as we were finishing our turn we were buzzed by two P-38s. They were very playful and made fighter passes at us, it was an exercise that was annoying at first but it meant the weather must be OK. If these guys could take off we should have no problem landing--let 'em have their fun.

(Results of the flash bomb photographing were a batch of overexposed negatives.)

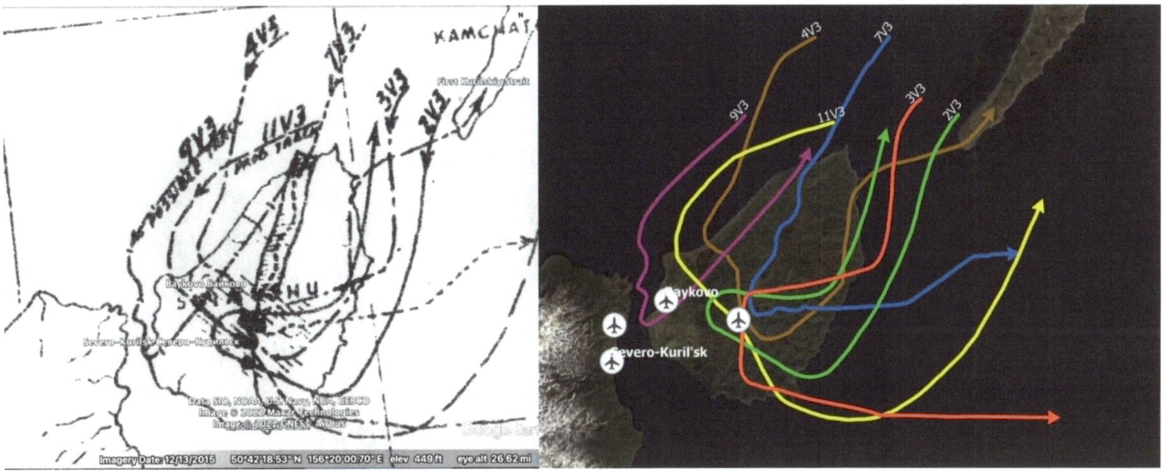

Photos 30, 31: WWII mission map of Sorties 49-225, 226, 227, 229, 230, 231 (left) and satellite imagery (right) with individual plane flightpaths.

Wings

It was excitement without containment as everyone scrambled to get in the car--my brother and I included. When we got to the cow pasture the pilots and several locals were already pushing the 'Jennie' to the farthest edge of the field. It looked like the whole town was turning out to see the show. Officially, this flight was probably logged as a training assignment that had to make an emergency landing. A crowd grew around the plane and my brother and I were lifted to see inside the open cockpits. The pilots wore their uniforms complete with the white silk scarf (a la Peanuts comic strip). After the plane was tied down and we started for the car the pilots asked my brother and me if we wanted to carry their parachutes? Wow, I was never so proud. "But don't grab the 'D' ring!"

Photo 32: Barnstormer Bill Lindley promotional photo. [Texas Barnstorming Museum]

The Glenn Curtis Model plane, a tractor design, had easily passed all the tests required by the Army Signal Corps and earned an illustrious record in the first World War for its speed and range. The first order for 72 planes in 1917 was soon doubled and eventually a total of 2,812 JN-4Ds were delivered. After Armistice was signed in November 1918, the Army Signal Corps had a glut of crated Jennies to sell or burn. At auction the public could buy a Jennie with engine for about $700 in the box. Then the stunt flyers and barnstormers took over and Americans went air-minded (and still are). At county fairs the people came to see the flying stunts, not the lions and tigers. The pilots would perform a perfect loop with trailing smoke. Diving down at top speed the pilot pulled up adding full power. As the plane reached the top of the loop the engine would cut out and he had to finish with a dead stick. The audience held their breath. This planned stunt was exciting for everyone including the pilot. The early planes used float-type carburetors which didn't work when turned upside down. Hopefully, the engine restarted. If it didn't, the air show was over. Barnstormers may have been the safer bet. After making several low passes over a town the pilot would land nearby and sell rides. Frequently, he was treated royally by the townsfolk to a free bed and breakfast. Not so for his hapless mechanic who usually spent the night fixing a flat tire, replacing worn spark plugs, and cleaning the passenger seat of vomit.

The world had changed--we had airmail at only 6 cents an ounce, Lindbergh had crossed the Atlantic in 34 hours, and the old war manuals of battle tactics were being rewritten. Another change came to youths like myself. Flying became my dream. Although I was reasonably accomplished with the violin at a high school level, my dreams changed to flying. However, that fiddle allowed me a college scholarship for two years, thanks to my Mom's patience and perseverance. I enjoyed my spare time making balsa wood airplane models, clipping air pictures, and reading about planes. I pasted aviation pictures on my walls and hung the little models of war birds from my bedroom ceiling; the Fokker D-7, Curtis Robin, WWI Spad, and (of course) a Jennie. Returning one day from San Antonio after delivering some produce, I decided to detour by the new Randolph Field and hope to glimpse the training activity from the highway. When I saw the main gate was open I simply went in. A grizzled old sergeant seated at a small table yelled, "Whatta ya here for"? I told him I wanted to learn to fly. He looked at me, dug out a pamphlet. "Here, read this." then turned his back and laughed.

Our parents held higher aspirations for their four children and set about making sure none of her flock would be chicken farmers. I was to be a violin virtuoso, my brother Morris a doctor, sister Nelda a music director, and beautiful Beth would marry rich. My Dad was one of eleven children, six boys, and five girls. After the WWI Armistice, he returned from France hoping to reclaim his job with the local bank where he had been a clerk, but to his chagrin one of his sisters had taken his old job and wouldn't give it up. No job guarantees in those days. But, Dad had another vision and with Army backpay, he started the town's first garage and filling station. The local Ford Model T and A owners no longer had to go eight miles to get serviced. In those days a silver dollar could buy five gallons of white gas with a nickel change.

Our mother had been writing to the registrar of a west Texas college (Abilene Christian College) about her talented boys. She eventually got a promise that the school would consider a one-year music scholarship for Bubba and me. Dad packed four of us into a Model A Ford and we were off to higher learning... For me, that first year of college was a near disaster--I almost flunked the mandatory freshman Bible course and barely passed everything else. Abilene Christian was like a divinity school--Preacher Boys got the 'A's. Except for math and chemistry, there were no engineering studies and dormitory life was the pits. The College Dean held a strict code of campus conduct; like, no dancing, smoking, early curfews, and 10-o'clock chapel every morning. This school wanted protestant priests, not engineers. Let's be honest, high school was so easy I didn't learn how to study. But, rumors about a new course in aviation was going around and if I could keep my grades up and pass the physical the government would pay for ground school classes plus flying lessons.

Photo 33: Abilene Christian College late 1940s [acu.edu]

The government-sponsored Civilian Pilot Training (CPT) course was offered in the next semester in my sophomore year. I dug in on my studies and my grades improved. My freshman math professor would teach the college ground school. Subjects included navigation, aerology, FAA regulations, mechanics, and combustion engines. For the first time in my college career my report card displayed an 'A' in a subject. My poor parents went wild with joy. Out of nine boys and one girl in the aviation class, I was one of four that would advance. After eight hours of instruction, I soloed in an Aeronca at the local airport. I was issued a flight certificate and it said that I was an aviator and they spelled my name correctly. The next flying course was to include a solo cross-country flight and learning aerobatics. Learning to solo was a thrill but, this next stage of flying proved to be exciting. Our flight instructor had purchased a tandem seat, single-wing, fixed gear, 4-cylinder Ryan ST with a wooden prop. It had that mean look of a fighter. The Air Force had used a version of this model with a radial five-cylinder engine, it had more power so their students didn't stall on takeoffs so much. The less powerful 4-cylinder Menasco engine became surplus and a bargain because the AF had rejected all of them from their cadet training. (Too dangerous, they claimed).

For my first flight in the low-wing Ryan, I was told to get a parachute, tighten the straps, wear helmet and goggles and climb in the rear cockpit. The Ryan plane proved to be fun though landing was a lot hotter and squirrelly. Its wing was stressed to be a stunt plane and could do loops, slow rolls, chandelles, snap rolls, and my favorite, the split S (spelled with an 'E' most of the time). My instructor let me do some slow turns and a practice stall or two then he put it into a spin which almost broke my neck. The Ryan recovered OK and so did I. Compared to the bi-wing Stearman it was fast and responsive. I had several solo hours doing aerobatics and enjoyed all of it. My favorite stunt was to start with a snap roll, when inverted do a complete half loop, then pull up for a chandelle ending at the original altitude and heading. My instructor didn't ride with me then and preferred watching my maneuvers from the ground with binoculars.

About this time I was halfway into my second year of college with poor grades in everything except the aviation classes. My twentieth birthday was coming up that November and it was predicted we would also be in the war soon. Volunteering for military service gave me more options than being drafted so I headed toward the Abilene Post Office where the Navy recruiting office was."You'll hear from the Navy soon."; said the Yeoman after he made sure all the forms were done "Maybe by Christmas";. I had applied for the Navy's V-5 program for aviation cadet training and was given a bus ticket to Dallas for more exams. Hello, the Japs had struck Pearl Harbor. We were in a war!

Photo 34: A U.S. Navy Ryan NR-1 training plane at the Naval Air Station Jacksonville, Florida, 1942.

Plane Jane

When a mechanic opened the floorboards of the Lockheed Ventura for inspection he spotted a note written in chalk on the main wing spar--"Good luck, guys. Hurry home." Signed "Jane". The mech let out a yell "Hey, Chief, come see what I found!". Soon the whole inspection crew scrambled to see the mysterious message from home and the news spread fast. Navy scuttlebutt spread and is known to be the fastest means of spreading information in the services, and it is wireless. But, could anyone guess who Jane was; wife, daughter, sweetheart, and was she pretty?

Not getting any mail was a chronic seaman's grouch. An excerpt from the U.S. Postal Pledge reads "Neither snow nor rain nor... ." As noble as the U. S. Postal pledge reads it was rarely the case for our mail planes from the states who white-knuckled it through the storms that often raged along the Aleutian chain. There could also be another cause of mail delay that happened from the start at the Fleet Post Office, San Francisco. Many FPO mail clerks had never heard of the Aleutian islands so they simply tossed your dear letter into the sacks going to Australia or New Guinea. If your sweetheart's letter smelled a little like coconut oil, that Jane's chalk note probably went south instead of north.

Our squadron's newer planes had search radar installed in our plexiglass nose. The antenna dish swept back and forth like your windshield wiper putting an orange smear on the scope. The scope was mounted under the radioman's table between his knees. The radioman could see to tell the pilots which way to steer towards or away from a target. The 11th Air Force planes were not equipped with radar at this time and relied on clear WX to spot their

Photo 35: April, 1943 - We reach Adak, AK and have first operational flights, sector searches.

targets. Lacking visual sightings this made the famous Norden bombsight useless in bad weather or at night. So, someone (not the Pentagon and certainly not MacArthur) figured out a way to use this Navy radar plane to lead an AF bombing mission over Kiska. Discussed and Practiced this plan was approved for a mission using a Navy PV-1 to be a bird dog and lead an 11th AF group. It would allow the AF to find and bomb the Japanese garrison on Kiska -- just two hundred fifty miles away. The Japs couldn't see us for the clouds, but we could find them with radar.

After the Japanese attack on Dutch Harbor our worst enemy, the weather, claimed more victories than Jap Zeros and anti-aircraft fire put together. This time weather would be our ally. Our search planes could see through clouds, could find surfaced subs, could locate landfalls and now we could lead a bombing raid and not be seen. We welcomed the fog and williwaws--the WX elements would be on our side. Who knows, if we bombed the Japs off our sovereign country, we might go home sooner.

One by one the bombers climbed through layers of clouds and emerged on top to find bright sunlight. A loose circle of planes now jockeying into their text-book 'V' formation. The final plane arriving was our Jane Plane with its radar. The Ventura took a westerly heading for Kiska Island and the bombers fell in behind. Being lighter and nearly 30 knots faster the Jane Plane was nearly out of sight in 15 minutes. Fortunately, a Navy crewman noticed this and warned the pilot to throttle back. Our ETE (estimated time en route) was 2 hours, ten minutes--enough time to make drop calculations and allow gunners to test-fire their guns. Navigation to the Kiska target was easy as the Kiska volcano jutted above the fog blanket. The aiming point was Little Kiska a small rocky lump that guarded the harbor entrance and installations.

Photo 36: Three PV-1 Venturas in formation near Adak, AK.

Radar began to form a picture of the rock and bay--bomb release was minutes away. A mike clicked, "Sir, I have the rock 5 degrees starboard". the pilot rogered and eased right rudder. Another click, "Sir, I'm painting--7,6,5,4,3,2,1,MARK," The co-pilot started his stopwatch and moved the bomb bay doors to 'OPEN'. The bomber formation watched and followed. The automatic pilot was engaged to hold a precise heading for another 22 seconds. This run could be a cakewalk.

But, so intent on holding course no one noticed that the Kiska volcano was becoming totally visible from its bottom to the caldera. We were losing our cover--the fog was dissipating--! Now the enemy could find us, AA could see us, and their fighters could come up. Escape? Some might escape, but the slower bombers would be picked off. A bombing formation cannot zig-zag or outrun fighters. We were in a 'pants down' dilemma.

The radios opened up from every plane with a barrage of cursing and praying. Everybody called everybody names and ugly threats at top of their voice. Our Jane Plane now became a Judas goat. However, no one reported seeing any fighters or flak. By a light signal the bombs were salvoed--the radio chatter eventually went silent as the formation slowly turned home unscathed. Finally, leaving the target area a bombardier reported he observed that the target area was hit. Nobody cheered. What happened?

Two weeks earlier under the cover of fog, a Japanese task force had slipped into the Kiska harbor and evacuated the garrison of 5,000 troops by using fast destroyers and their giant I-400 submarines. When we staged our forces to retake the island the only living things left by the enemy were a few mangey dogs and some chickens. Oh, yes, even though the U.S. knew the island was not defended we still went ahead with a full-scale invasion force and retook our island. It was reclassified as a military exercise. There were some casualties, mostly from booby traps and frostbite--the mangey dogs were shot.

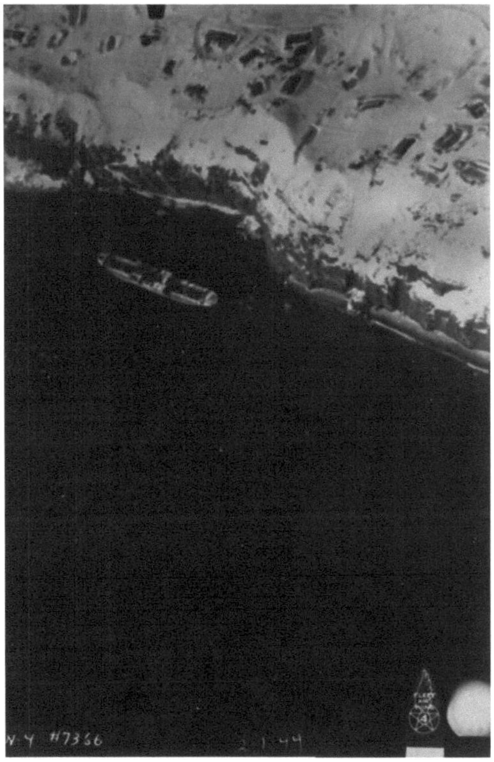

Photo 37, 38: Grounded Japanese freighters at Kiska.

UFO Sightings

Have you ever seen a real UFO? I have. Most of the stories and pictures describe them like pinpoints of light that streak across the night sky. There are also the close-ups where a saucer-like object alights and little green men emerge and converse with the locals in English. Most accounts are by a single witness and some take actual photos. My sighting was in WWII when I was a pilot on a night bombing mission to photograph and bomb a target in the northern islands of Japan. We were about halfway there and I had set and flipped the autopilot on as the weather was relatively smooth. The outside air temperature at our altitude was near zero degrees. The outside navigation lights were off and only our infra-red cockpit lights were used to see the gauges. Overhead the stars, usually brilliant, were partially blurred by some high cirrus clouds. Our navigator would have some trouble getting a good fix on Polaris and Betelgeuse. This was not a serious concern as we can still detect a faint signal from the homing beacon at Attu. It was time to test fire the .50-cal guns and the turret gunner was chinning himself into his seat, there are no steps to help. When all stations are ready we will test-fire a couple of bursts.

Suddenly, I get a tap on the shoulder--the navigator whispers for me to look back out my window because the crew had spotted something. I look and sure enough, I see an unusual reddish/orange streak-like fire behind and a little below us. It appeared to be about a mile away, but hard to tell. My first thought was that it was another plane's exhaust flare as there were other squadron planes on this mission. However, we were all familiar with our own flares and this was different. Who else is in mid-ocean? I announced to the gunners to keep it in sight, but not to fire. Minutes passed and the object grew bigger. Fingers were on the triggers. The mysterious flare of light was now becoming saucer-like. What or who was it?

Then we saw it clearly. The low cloud skud that forms over the sea was giving way to the big orange moon rising in the east. Now it changed into a beautiful golden moon rise. Gee, we almost shot it down.

Photo 39: 1943 - Two by two we head to the Aleutians. Our first stop was Ketchikan.

Photo 40: 1943 - Adak - Consolidated PBY-5A Flying Boat and Lockheed PV-1 Venturas.

The morning of March 25, 1943, was crisp and clear, but it was roaring noisy on our tarmac. Our squadron of 15 warplanes and their crews were getting ready to leave the USA and head for the Territory of Alaska. The first 3-plane group was airborne by 10 AM with new planes, new crews, and a new purpose--to dislodge the enemy's occupation of two islands in the Aleutians. The squadron would head from our training base at Whidbey Island and fly over Puget Sound to the Pacific Ocean, then Alaska. The first stop would be Annette Island. From then on everything would be a unique adventure for all of us. I was the co-pilot/navigator of a Navy PV-1 Ventura made by Lockheed. I swapped flight responsibility with Ensign Cornehl who also trained as an Ensign Co-pilot/Navigator.

Our PPC Lt. (j.g.) Bermingham) was the only one of our six-man crew who had prior experience flying in the Aleutian Islands. In the first few weeks of the war Berm had been a Catalina co-pilot in a PBY squadron and then was reassigned to this new group designated VB-135. Soft-spoken and likable, Berm was our guide and mentor. Like him, our squadron acquired several more veteran P-boat pilots rotated from Alaska. At first, some of these veterans were reluctant to make a switch, they felt comfortable in the old, slow, stable, yoke boats. But, most younger generation of pilots (like me), went for the newer and faster fighter characteristics of the Ventura. We did not want to be sitting ducks. That is what had happened to the old PBY squadrons, they had become sitting ducks for the Jap fighters. This was a drastic military change for our Navy and patrolling our shores went from lumbering float planes to a fast patrol bomber that could defend itself against enemy attacks.

We landed in 3.7 hours touching down on a long graveled strip at Annette Island, Alaska. Canvas-topped Army trucks picked us up and took us to a building that was filled with folding cots and had one wood stove. Officers occupied one side of an auditorium and enlisted men took the other. Immediately a fight among the crews started. One of our senior officers tried to quiet the rowdiness, but got little results--the scrapping continued and more flights and their crews arrived tired and hungry What's more the wood for the stove ran out and by midnight it was very cold on the bare cots. Many of us were warned to bring our heavy flight gear to sleep in. Some didn't.

"When do we eat?" The cry went up. The Navy personnel had nearly doubled the number of mouths to feed. "Sarge", the Army cook, went ape when he saw the Navy guys lined up for chow. He had to feed them and finally served that old standby "Shit on a Shingle". For our breakfast, we had powdered eggs with powdered milk. By the time we left the Army cupboards were bare. One evening we were greeted and invited to a basketball game by the neighboring village of Metlakatla. Transportation was provided to go to their high school auditorium and a school bus took us to their gym. Since weather had delayed our flight departures we had little to do but sit and wait--it was a welcome relief to get some exercise for a change. Luckily, several of our pilots had played college basketball and were eager to take on the native Athabascan tribe. All the seats were removed from the auditorium floor so there was just enough room to line the boundaries for a basketball court. One basket was hung over the stage and the other basket was nailed just above the entrance door. The sidelines were already filled by wives and children allowing just enough room for a single person to stand. The women and children all cheered and chanted loudly and appeared to favor our team. When the teams entered the ladies booed the native team and cheered the Navy. The native team, however, did appear in uniforms--their players all wore rubber boots and stood about five feet tall. At first our Navy team looked to be favored--but that changed. The 'A's stomped, passed, and shot with unerring skill--the final score was a lopsided embarrassment for the Navy.

After four days of waiting for a break in the weather, it finally cleared enough for some planes to leave. It also allowed a couple of Canadian P-40s to try to take off for their routine patrols. Then, to show off for our benefit two Canadian planes lined up together and attempted a slow roll on takeoff; one rolled to the left and the other to the right. The P-40 on the right didn't finish its roll and crashed. We wondered if this was some kind of omen?

Photo 41: Snowcapped peaks block take offs and landings to the north and west runways. Gehres Point where pilots and staff live forms eastern border at Massacre Bay. "Big" Clark's crash was in center of Massacre Bay. All perished. The plane was carrying magnesium flash bombs for night photo reconnaisance.

It took six days for all planes to get to Kodiak Naval Air Station--our halfway mark. Kodiak NAS was a welcome change. Their facilities were first class with comfortable good sleeping quarters and the best bar on the west coast (so they say). Also, we had a chop or steak every night at the BOQ, two to a room with showers, and maybe some nightlife in town. A few of us were eager to see this fa-

Photo 42: PBY-5A Crash on Attu "Fierce Willie" was destroyed.

mous frontier town, so four of us piled in a cab that was idling at the main gate and off we went to see the business section of Kodiak. Maybe, to a smelly trapper just in from months in the wilderness this mud hole might seem like his heaven. But not to us. The town of Kodiak was mud streets, wood sidewalks and every other business was a bar and the barman was a pimp. Military personnel was scarce and probably preferred to stay at their base, no wonder.

Patwing Four at Adak was headquarters for the Navy and 11th Air Force. It was within striking distance of Kiska and Attu now occupied by the Japs. Commodore Gheres was the Navy CO here and gave us a nice welcome. He lost no time and immediately assigned us to the usual sector and reconnaissance patrols. On May 4, 1943, the squadron advanced closer to Jap-held Kiska to a new island base on Amchitka Island. On May 12, 'Dog' day for the invasion of Attu, our crew logged 6.2 hours covering our invasion fleet which was still unloading men and equipment on the beaches at Attu. Some Japanese bombers were reported heading our way, but the weather and the 11th Air Force fighters forced the Jap 'Betty' bombers to turn back to their Kurile base. The Battle for Attu which had been planned for only a one-week engagement lasted nearly a bloody month. It was one of the coldest and costliest battles of the Pacific. It finally ended after the enemy tried a suicide banzai attack with about 400 men. These remaining stragglers were eventually wiped out but not before they had butchered nearly all the patients of an Army Hospital in their final 'do or die' attack.

After May 1943, with hardly more than two months of flying time in Alaska as Co-pilot/navigators, six senior Ensigns of VB-135 were considered 'experienced Aleutian pilots' and were sent back to the states to start a new Ventura squadron. The new squadron was designated VPB-135. Meanwhile, in the Central Pacific the United States was beginning to take the offensive.

More training began in PBY-5As at Whidbey Island with an emphasis on instrument flying. As Co-pilot /Navigators none of us had hardly any hands-on experience as pilots in the PV-1 Ventura. But, this would change. More PV-1s arrived so we were able to use them for instrument work and no longer be a 'sitting duck' plane. We would not return to the Aleutians in PBYs, but in a new fast and deadly squadron ready to go with 18 crews, 15 planes. Most of the new Pilots had come from their former duties as flight instructors at Pensacola or Corpus Christi. Some of us discovered an old instructor we knew from cadet flight instruction days. But, whether they were assigned or ordered to the Aleutians was not discussed. Old and new pilots learned from each other. Senior officers had first options for aircraft, crews, or equipment, but we were all trained together. The PV-1 had new radar and the latest low approach, radio communications. We were up to date and equal to the senior pilots in our experience with the faster Venturas.

In 1944 my assigned co-pilot and navigator were also well-trained pilots with a long period of flight school. Each wore the gold wings and wanted to show their piloting ability. My co-pilot, Kay Walsh, from Kansas City, had been a swimming champion in college and was good enough to be on the US Olympic team for the next world Olympics. But, Hitler had put a stop to that. Dick Rice, from Milwaukee, was a typical college playboy. A handsome six-footer who probably never cracked a book for whatever 'mysterious' degree he sought, but somehow he sailed along with reasonable college grades. Our crew liked him but didn't know if he, as our navigator, actually knew where we were some of the time. Rice only took a pencil and blank chart paper aboard for each mission. The crew worried about how he could steer us 1400 miles over the deep, dark waters of the Bering Sea. So, to build more confidence for the crew I took Rice aside and suggested he bring a few more instruments with him on our next strike--it would help reassure the crew that he actually had navigational skill.

Photo 43: Pilot Kay Walsh #18 crew in 1944.

Photo 44: 1945 - Attu - Dick Rice at Jap dugout.

Fighter Intercepts

Photo 45: Pat in the cockpit.

On my 2nd deployment to the Aleutians in 1944, I flew 25 missions to the targets as a pilot of a Lockheed PV-1--the earlier version of the Harpoon. Taigh Ramey and his very capable and dedicated helpers have restored the PV-2 and allowed me to take the controls again after over 70 years. It has revived some of my memories.

As a pilot I was assigned to a Navy Patrol Bombing Squadron, made up of 18 flight crews...in the Service there is a pecking order according to your rank and serial number -- My crew was number 18...our skipper was #1, executive officer #2, etc. I was the lowest-ranking PPC of the squadron and was not allotted any of the 15 planes...I always had to fly in another pilot's plane...his name was stenciled on the fuselage. However, I had the best crew and told them so--later we proved to be so.

Each crew trained and flew together; 2 pilots, navigator, radioman (radar operator), plane captain (mechanic), and turret gunner. The mechanic took over the tail gunner spot in the air.

We were based at Attu, Alaska, the westernmost territory of Alaska. The International Date Line zig-zagged around Attu. Uniquely, we joked that when we took off for the target we would arrive there tomorrow, and returned back to our base yesterday.

About a year earlier Attu was the scene of one of the bloodiest battles in the Pacific... our troops finally wrested this cold, bleak, uninhabited volcanic rock from the Japanese garrison. Instead of a 3-day skirmish, it took nearly a month of freezing hands and feet to do so. Why not just leave this island alone--the enemy would certainly perish due to the elements alone.? Who, but the native Aleuts could survive here? I was a co-pilot then and we flew fighter cover for the task force troop landings on D-day Attu. We didn't engage any fighters due to the ferocious weather, but that's another story.

Strategically, Attu was critical for the plans of invasion... it would become the northern invasion front. The Pacific War, despite MacArthur, was beginning to favor our forces. The Battle of Midway gave us the edge, and plans were to invade Japan--from the south, east, or north. To defend their homeland the enemy began building their defenses in the Kurils, especially the two northern islands near Russia (Paramushiro and Shumushu).

Soon as the Seabees finished a steel runway the big 11th Air Force transferred their squadrons to Attu. They had heavy bombers; B-24s, and B-25s, and fighters P-38s and P-40s. It was a 10-hour trip for the bombers from the Aleutians to the Kurils and our fighters were incapable of giving them any protection--too far. Anyway, a big raid was planned to attack the facilities at Kakamubetsu--.

With full bomb loads in 8 B-24s and 12 B-25s, the formations approached their target. The Japs were ready; Fighters were already up and struck first on the initial bomb run, AA had them dialed in when they got over the target, and the fighters picked off the stragglers leaving the target area. One B-24 and 2 B-25s were shot down. Another B-24 and 5 B-25s were so badly battered that they barely made it to neutral (Russia) territory. Of those who made it back to Attu, 2 pilots, 1 co-pilot, and 4 crewmen were badly wounded.

It took 6 months for the 11th Bombardment to regroup, retrain, and assume their bombing role. Morale was the pits.

Meanwhile, VPB-135 and VPB-136 were challenged (actually we were ordered) to get photographic data of any military value. In order to avoid fighter's night photoflash bombing was done. The results were rarely helpful--clouds, difficult to handle and drop the magnesium flash bombs, blinded by searchlights which often tripped the photo sensors, and locating a target above the clouds by radar was difficult. However, no enemy fighters were up at night, and we avoided AA by changing course every few minutes.

By some luck one of our pilots (Vivian) was on a sector search that was aimed to get target WX before the next mission--a mission to pinpoint an unknown fighter base on Shumushu. The WX was CAVU and the pilot spotted the new fighter strip and radioed the base for permission to fly inland for photographs. Without waiting for an official OK the crew continued inland anyway and photographed the base. Then dove and high-tailed it back on the deck outrunning the fighters, Vivian and crew were safely on the way home when the permission was finally granted.

This chase changed our mission strategy--no more night photoflash bombing--we could get pictures and make bomb runs in the daytime. And, had a good chance of escaping by diving or finding cloud cover. Not counting routine sector searches for submarines, night photo, and familiarization flights, I had 8 interceptions by Zeros (the Mitsubishi A6M) referred to as a Zeke.

On a daylight mission (8/25/1944) to bomb and photograph Kurabu Zaki, we were jumped by fighters who had been hiding in high cirrus clouds. When over enemy land we always kept close watch for fighters--the navigator (Ens. Rice) was looking aft from the astro dome bubble and so was the turret gunner. We were alone and had made a shallow dive to release our 3 bombs. A bogey was reported out of range port side and all eyes were watching for an attack. The next thing I knew was a yell on the intercom that Rice was hit and down. A 20mm shell had struck the antenna post and exploded. A fragment came through the plexiglass astro dome and hit him behind his ear. A head wound bleeds a lot and this one bled a lot. I nosed over into a power dive and headed for some low clouds near the deck. In minutes we reached a safe distance in the low fog at sea. We were chased out to sea. The crew in the cabin finally stopped the bleeding with sulfa powder and I was told the wound had been plugged. I suggested we may have to go to Russia only about 1 hour away, but the crew all went thumbs down.

On another daylight photo run my crew was assigned a special rigged Ventura. An extra fuel cell was added in the bomb bay and Fairchild long-range cameras were mounted in the nose and tail. Nose cameras had one vertical and two obliques at 30-degree angles. Tail cameras replaced the guns; one vertical and one shooting 30 deg. aft. Also, a handheld K-20 camera and a photographer's mate added to the crew. We were to make a run from north Shumushu to south Paramushiro photographing 5 installations We took off at 0100 and followed 6 other planes which were to divert fighters from us by bombing and strafing the big fighter field at Mioshino.

Our diversionary forces' surprise came off successfully, and we were all surprised at the beehive of excitement on the field. You could see fighters taxiing and some taking off. and many still idle in their revetments. "It was CAVU--not a cloud in the sky. AT 11,000 ft. we headed for our first target, Kataoka. We had a bomb bay full of extra fuel and could set up for speed. Everyone craned for bogeys, but none appeared. All 5 cameras were turned on as we reached the first field, then we spotted the second target and the crew shouted they saw more planes taking off below. We continued south and over Kashawabara when two bogeys appeared ahead of us and went by before they knew who we were. Our next target was Dago Zaki about 40 miles south. The two Zeros we passed turned around and gradually began overtaking us but keeping a safe distance from our top turret guns. Jacobsen gave them a burst when they got too close. Having a healthy respect for our twin .50s the fighters quickly jinked out of range. After a few minutes of this, they stayed parallel to us and then climbed for the typical fighter passes at 4 o'clock and 8'oclock high. My navigator stood in the astro hatch, ready with the intercom mike and warned me when one started his run. All I had to do was turn in direction of the fighter's pass and this shortened the angle so steeply that their fire always went behind us. The fighters were faster and more maneuverable than we were at this altitude, but our relative speed was the equalizing factor. By the time we reached Dago Zaki we were untouched and felt we had some kind of combat edge.

By now we were coming up on our final target, Dago Zaki, and discovered several fighters had spotted us. They had come up from the big base at DZ. Everyone was shouting and jamming the intercom. Amidst all this our turret gunner, Jacobsen, got a 'probable'--he certainly had plenty to shoot at. Now, two things were running out; the film magazines and our cool. With the small K-20 camera we managed a window shot of Karabu Zaki. Miraculously, we still had no damage. I made a shallow dive from 8,000' to the deck and firewalled everything. We were followed by 3 Zeros, but they only feinted runs and the turret gunner kept them out of range. On the way back to base I learned that the crew had threatened the Photog-Mate with lynching if any of the cameras failed. To everyone's relief (especially the Photog) we had over 350 pictures of excellent quality. I also learned something else from a writer/correspondent/and friend that this fighter trio included a Japanese Ace with 87 'kills'.

Two of the planes of the diversion group had been hit, one lost an engine, but made it safely to Russia. The other four planes made it back to base but 2 needed engine changes for continued use of full power; two other planes received AA skin damage.

Later, to photograph an update on the completion of the new strip at Kakumabetsu we were chased by 11 fighters separately or by twos, Most kept a respectful distance for a while as if checking us out. Two A6M Zeros came in too close and our turret gunner hit both. He was credited with a "kill" and a "probable" We sustained minor skin damage in the tail.

Respect for the Lockheeds grew! But, also rumor was growing about the Kamikazes.

Photo 46: VPB-135 PV-1 flying near Adak, 1944.

The Admiral's Cow

No story of the PV squadron's exploits in the Aleutian Islands would be complete without the story of the Admiral's cow. After the capture of Attu, the Wing Commander of Fleet Airwing Four decided to move his headquarters from Adak to Attu. He ordered the SeaBees to build a suitable Headquarters for himself and his staff adjacent to the main airstrip on Murder Point, Attu. The ensuing structure was referred to as 'the palace' because it had many amenities such as hot running water, a field telephone, etc. The Admiral decided another of the missing home comforts needed for their drab existence was fresh milk. He and several thousand other WWII servicemen did not like the dried milk of that period. It would not stay in solution. Stir as you would, before you could drink it, it had settled to the bottom of the glass where it looked, and tasted like chalk. It was decided that the solution to the fresh milk problem was to get a cow. Now, cows were as scarce as Humming Bird's nests in the Aleutians...the native Aleuts had never seen one. The treeless terrain, cold weather, and lack of suitable forage were factors. However, through the 'can do' ingenuity of the U S SeaBees any obstacle can be overcome and this bossy would have a palatial heated home.

So happens the Seaplane Tender USS AVOCET (AVP-4) that was attached to FAW-4 was due for an overhaul in the states. The Admiral solicited contributions from members of his staff and gave the funds to a member Aide boarding the AVOCET when she sailed for Seattle. The lucky Aide had a farm boy background and orders to purchase a GOOD cow. This trip was the answer to the prayer for the ship's crew who had been at sea ever since the war started. Now the crew would have stateside liberty and GIRLS--week after weeks. But all good times end sooner than expected and the strange-looking Man-0f-War started back to Attu with a cow tied to the fantail and the hangar full of hay.

While the AVOCET was on her mission to Seattle, the Admiral had the SeaBees build a barn and included some chickens to help the cow feel at home, but the barn had to be located far enough away so odors didn't permeate the headquarters area. Its fence was staked out next to our living area and we got a healthy whiff of the cow pen every day on our way going to chow, briefings, and the wardroom. The Aide had indeed purchased a fresh cow and there was more milk than the Admiral's staff could use. So, in good faith, Headquarters donated the extra milk free of charge to the Officer's Mess. A tall glass could be purchased at the Wardroom bar for ten cents. The taste of fresh raw milk is different from the homogenized variety and many men didn't like it. Others, seeing and smelling this menagerie questioned the sanitary conditions. Would the milk pass for Grade A?

Photo 47: Cow barn on Gheres Point, Attu with attendant and chickens.

But, at last, a squadron member came up with a solution--Alexanders! The bar could use the milk to make fresh Alexanders, reasoning that the alcohol would take care of any germs. But, do the math, old Bossy could produce nearly 2 gallons of milk, and a tall Alexander uses about 2 ounces. The cow and her entourage were not long for this world. Early one morning those close to the barnyard were awakened by a bugler sounding 'Tatoo'. When they poked their heads out there was the Admiral's staff resplendent in full dress uniforms drawn up in formation facing the barn. One of the Aides was reading the orders transferring the cow from Fleet Air Wing Four to an Army Engineering outfit. A deal was made and the Army outfit had fresh steaks and hamburgers for their staff.

Photo 48: A lady cow arrives on Attu.

Kodiak

Kodiak Island was our stopover on the way to the Aleutian Islands. Also, it is where civilization comes to a muddy halt. Further land travel north or west is by sled dogs. The Japs had already taken part of our Alaska by simply walking ashore without any resistance. Further invasion of North America would have been easy if Japan had not lost its carrier fleet at Midway. For a short time, Alaska remained practically undefended. Except for a Naval Air Station, the Kodiak we saw was still just a pioneer village that existed for the benefit of trophy bear hunters and commercial salmon fishermen. The town streets were unpaved, with wooden sidewalks and no street lights. Every other business in town was a tavern and every other tavern had another business run by an enterprising madam.

We flyers had no business going to see the town except curiosity. Few cars, no traffic, and no place to go except the shuttle back and forth to the NAS. But, we struck out to see it all. A couple of beat-up autos idled at the Navy gate with a windshield placard that read 'Taxi'. For two bucks a head they bounced you in to see the sights. Four of us hailed a cab and took off. For fun we threw some questions at our driver like: Where's the best nightclub? any gay bars? Can we get a bear-burger? Got any suggestions? He answered, "Well, my wife can help there", he said,, "I don't think she's busy tonight." We didn't ask any more questions.

Photo 49: Main Street Sears and Roebuck Kodiak

Photo 50: Bank of America Kodiak Alaska postcard.

There were a few businesses on the main street; a Post Office, hardware store, assay office, and a Sears Roebuck mail order store. Every other tin-roofed building was a saloon with a neon sign flashing 'Welcome'. Our driver let us off at one with this warning: "Don't believe the label sez what's in the bottle." Except for a little space at the bar the place was filled and humming. The bar was circular and you could watch everyone. Two ladies were seated one at each end of the bar, One had a big smile laughing loudly and trading jokes with her audience--the other lady was hunched over her drink muttering to herself. . The happy lady was asked if she knew what was upsetting her sad friend. "Yep", she said, "You see my new teeth"? she showed a wide grin, "Got 'em yesterday, but Maggie's ship ain't arrived yet".

Set Your Watches

Photo 51: Copilot Kay Walsh in the cockpit.

My copilot, Kay Walsh, always brought his .38 caliber Smith & Wesson to the mission briefings. Our sidearms were standard issue for each six-man crew member but hardly needed at the pre-flight meetings. Kay was more like the silent type, a very good pilot and I was confident to turn the controls over to him anytime. Over 6 feet tall, slim, and trim he had been a student at Kansas City training for the 1940 Summer Olympics swimming team when war broke out. Of course, the Olympics was called off, but Kay kept to his training routine--didn't drink, smoke, or gamble and into bed early each night.

Carrying a sidearm on a mission to the target was optional and in my opinion, senseless. I left my pistol at the base as many others did because we had learned more of the Jap's treatment of their POWs--especially of our captured airmen. But, Kay always kept his gun holstered under his left arm.

Our briefings always followed a format that started with the Senior Wing weatherman unrolling his charts. "Well, guys," he'd say, "You might hit a little line squall two hours out and there could be another a little later, but the target should be fairly clear for you. No problem there." "Ha, ha" he laughed at his own joke." "Then, when you get back our fog should begin to lift about noon." (We were due back about 0800). Any questions? Our next briefer was not so jocular when we got target assignments, new call signs, code books, plane assignments, and the pills (those to keep us alert and those to make us sleepy). And lastly, a very accurate navigation chronometer. This mission was to be a night photo run to find any new defense gun emplacements. Except for bad weather, AA not fighters would be our biggest concern. When the blinding searchlights locked on they followed until you were out of range.

One night we were hit by AA flak and our navigator was wounded in the back of his head. Head wounds bleed a lot and this one bled a lot. We didn't worry about the AA or fighters now, it was how can we help the navigator before he bleeds to death. The situation was critical--should we head back to base which would take at least 4 and 1/2 hours, or go to neutral Russia only one hour away. I wanted a second opinion. I explained it to Kay as we got out of range and he said nothing for a minute or so, then he squared in his seat, unzipped his flight jacket exposing his pistol butt--"I'm not going to Russia!" And we didn't.

Japanese Ace Hiroyoshi Nishizawa

Photo 52: Hiroyoshi Nishizawa [Wikipedia]

It was a triumphant day for the bombers and they circled to return to their base at Rabaul. Their fighters had safely escorted them to the target and now the same escort covered them to return safely home. The bomb run had been a textbook exercise and our airfield at Port Moresby was peppered. The defenders were a few old P-40s and P-39s that managed to scramble and take off from their rough and muddy field. Our radar had not found the attack soon enough and our fighters were too slow, too little and too late. The Japanese fighters easily struck most of them down. In the South Pacific we were taking a licking in the air, on the ground, and on the sea. It wasn't until the battle of Midway that we found victories. The only way we could win in the air war was to get the jump from above, dive, hit, and run (like hell).

In the Solomon Islands, a Japanese Navy Air Service group had occupied a base at Lae, New Guinea, which had several air aces like Sakai, Ota, Nishizawa, and Takasuka. They were all experienced and had lots of kill skill. But Nishizawa was top gun. He had distinguished himself long before joining the Lae air group which was equipped with the more modern planes. Now, he was to fly one of the best, the Mitsubishi A6M Zero, the fastest, and most aerobatic aircraft of the day. As his record of victories increased two other aces asked him to join their three-plane team. With them he showed more tactful skill and leadership than ever. Some records gave him 102 kills, but 82 is the more accepted number today. Even so, it exceeded all other records. For example, he was credited with shooting down a PBY patrol plane, but that crew managed to get back to their base on one engine.

The night before the raid while listening to an Australian radio program it played a piano recording of 'Danse Macabre' by the Camille Saint Saens, also known as the Skeleton Dance or Dance of Death. Nishizawa was entranced. With him were the two other aces, and he asked: "You know the mission we have tomorrow? Why don't we perform a little show, a dance of death of our own? We can do a few demonstration loops

Photo 53: Nishizawa in his Mitsubishi Zero A6M3a Model 22, tail code UI-105, May 7, 1943. [Wikipedia]

right over the field, this should drive them crazy on the ground."

They were enthralled with the daring stunt and agreed.

The next day of the bombing three planes peeled out of the formation and headed back to Port Moresby. Then, over the airfield they formed a 'V' so tight the wings overlapped and flew three successive loops. Then dove lower and did another series. The AA trained their guns on the trio, but no one fired. That night the trio of pilots couldn't sleep for laughing. However, early next morning they were summoned to the commander's office and he was enraged. "See this," he screamed, waiving a letter in hand. "It was dropped earlier on our field".

The letter read: "To the Lae Commander;

"We were much impressed with those three pilots who visited us today, and we all liked the loops they flew over our field. It was quite an exhibition. We would appreciate it if the same pilots returned here again, each wearing a green muffler around their neck. We are sorry we could not give them better attention on their last trip, but we will see to it that the next time they will receive an all-out welcome from us."

The Commander dressed them down severely as idiots. and each tried to smother his laugh.

Photo 54: Hiroyoshi Nishizawa in his Mitsubishi Zero A6M3 Model 22 (tail code UI-105) from the 251st Kōkūtai over the Solomon Islands in May 1943. The unit's aircraft have been hastily sprayed with dark green camouflage paint on the upper surfaces.

Photo 55: I fly again in PV-2 "Harpoon."

My son, Lea, and I go to the air shows every chance to watch the air races. We especially enjoy seeing the old warbirds that are modified to race the large pylon course at Stead Air Base. There is an unlimited racing class that uses many of the WWII warbirds like the Air Force P-51s, Navy F6Fs, British Seahawks, and others. Their wings have been shortened, all with four-blade propellers, and special changes to the engines that increase their horsepower. These beasts roar around the long course a few feet off the ground and shave close to the pylons at over 400 MPH. There are other exhibitions at this show like the stunt planes and the crowd-pleasing military formations done by the Air Force Thunderbirds and Navy Blue Angels. But, the 'Unlimiteds' are the most thrilling.

One hot summer day Lea and I were sitting in the bleachers when Lea spotted a dark blue twin-engine craft in the parked exhibit area. Lea headed to the static area to check this old bomber out. He found a beautiful, refurbished PV-2 bomber such as I flew in the Aleutians. He returned all aglow and said we were invited to visit, and sit in the shade of the wings and have a cold drink at the owner's plane. Taigh Ramey, the owner, had spent a fortune rebuilding this Lockheed PV-2 that was found in a pasture near Napa. It had been used as a crop duster. We spent the afternoon examining the old Harpoon and I marveled at the care and expertise he and his crew had done at their Stockton hangar. This Harpoon had the upper ball turret with its twin .50-cal machine guns, five fixed .50s in the nose, and four rocket launchers under each wing. Inside was the new (at the time) Loran electronic navigation system. No more celestial star fixes. We spent the afternoon exchanging information. One of Taigh's ambitions was to find Navy veterans who had flown this plane in combat.

Then he could form a unit and go to other shows and exhibits around the country. So far, only myself and two other vets had been located in this area. Taigh's next air show was to be at Travis AFB near Vacaville. He asked me to go with him and fly in his Harpoon...wow, three days in Reno, all expenses.

When about 5 miles away the Travis control tower cleared us to land and added we could buzz the airport if we cared to. For the short 20-minute flight from Stockton I was still in control. We were on a west heading which put us aligned with one of the 2-mile-long runways. Taigh hand signaled me to make a dive and do a dust job. I took the plane down to 200 feet and Taigh was signaling me to get lower. It was a very safe dusting and I knew Taigh was not happy. He signaled me to pull up for a left downwind landing pattern--I did and noticed he had engaged someone on his radio. So, I set up for the landing; fuel selector on main tanks, landing gear down, flaps down, props high pitch, cowling and oil cooler flaps set, fuel pumps on high, mixture auto rich. We were trimmed and ready to make a controlled, descending landing. When properly executed the aircraft is to U-turn, descend, and hit the runway underneath, bingo, like the textbook says.

But, when we reached the letdown position I looked at Taigh to let him take the controls. Now he is looking out his window and didn't notice my signal. I went ahead with the landing checkoff and again tried to get his attention. He was now looking away and talking to someone--I was ignored. I thought: "do you expect me to land this thing--I haven't landed a PV-2 in over 70 years! We have a 15-ton multimillion-dollar antique bird, four passengers in the cabin--let's not begin this air show by strewing tiny pieces of aluminum and body parts around the airport everywhere".

I realized then it was too late to swap controls. I had to set it down. We touched down and heard the tires squawk. It was textbook.

Photo 56: Starboard engine oil leak caused emergency landing at Chino. The PV-2 was scheduled to be taken aboard the USS Essex Carrier for participation in air show at Honolulu.

Secret Aerial Weapon

Our Allied Air Forces used a secret weapon that was so hush-hush it was not revealed until WWII hostilities had ceased. Even now there is scarce reference in official reports of their effect on the enemy. In my home library I have the classified records of all bombing sorties of my squadron VPB-135. These highly classified reports are on microfilm and I have a copy of that film for every mission the squadron made against the Japanese Kurile islands--they include the official 'Diary' and the 'Aircraft Action Report of every strike. My records also include the details of the bomb load; i.e. incendiary, fragmentation, general purpose, and the .50 cal. gun ammunition load, cameras, and any special equipment. At the time all these reports were highly classified. But, they do not mention the secret weapons!

Delicate handling by a skilled crewman was necessary as one by one each weapon was carefully dropped through the open flare chute.. The navigator signaled with a raised hand when he determined we were over enemy territory. Usually, the drop was when a fog had totally covered the target area and made normal bomb runs impossible. Night was the best time as the fighters and searchlights couldn't locate us. Of course neither could anybody observe the effect of the drop, but the satisfaction of our crew was immense. A direct hit was not necessary, but we never knew. Anyway, accuracy was not essential--and a near miss could have the desired effect.

Just inside the planes' cabin door and nearly under the turret gunner's station is our head (or toilet). To use the head each visitor must place a special paper sack inside it as you do your garbage bag.at home. After use, the sack is securely tied and later dropped through the flare chute. Don't do this over Seattle. Anyway, this can be a unique missile and is called a blivet. Frequently we had blivet donations from our ground crews at Attu and they were welcomed and carefully stored for the next trip to a target. A blivet is actually defined in the Wikipedia as 10 lbs. of shit in a 5 lb. paper bag. (Look it up).

Can you think of a better way to complete your mission?

Airplanes I Have Flown and Pilot's License

The airplanes I have flown:

Aeronca L-3B Grasshopper
Beechcraft SNB-1 Kansan
Boeing-Stearman N2S-5 (Model 75) Kaydet "Yellow Peril"
Consolidated PBY-3, -5, -5A, -5B Catalina
Howard NH-1 instrument trainer, GH-1 Nightingale
Lockheed PV-1 Ventura
Lockheed PV-2 Harpoon
Mooney M20F Executive
Naval Aircraft Factory N3N
Navy Fleet Trainer
North American Aviation SNJ (T-6) Texan
OS2U-2, -3
Piper Cub J-3
Ryan ST Sport Trainer
Vultee SNV-1 (BT-13) Valiant

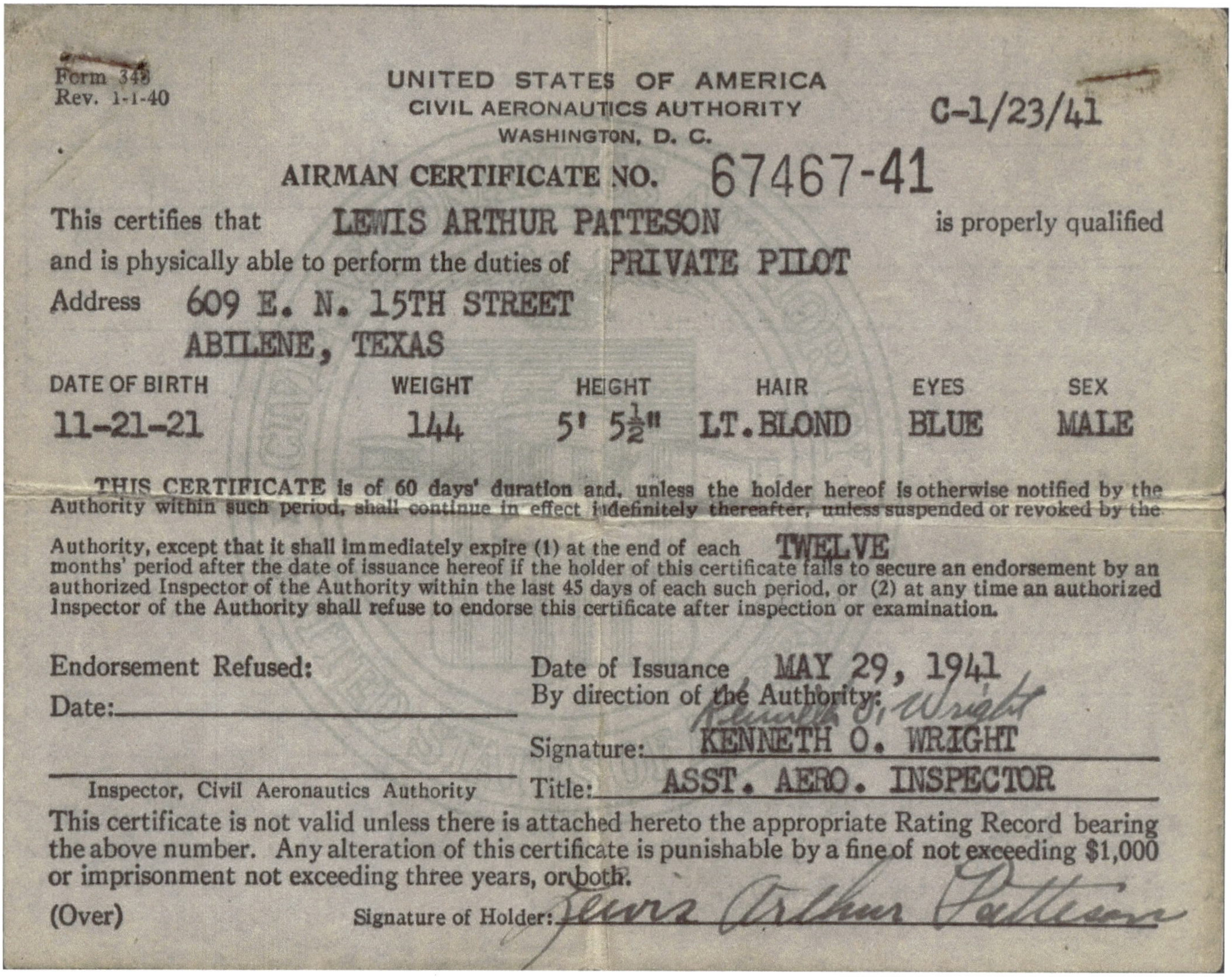

Form 348
Rev. 1-1-40

UNITED STATES OF AMERICA
CIVIL AERONAUTICS AUTHORITY
WASHINGTON, D. C.

C-1/23/41

AIRMAN CERTIFICATE NO. 67467-41

This certifies that LEWIS ARTHUR PATTESON is properly qualified and is physically able to perform the duties of PRIVATE PILOT

Address 609 E. N. 15TH STREET
ABILENE, TEXAS

DATE OF BIRTH	WEIGHT	HEIGHT	HAIR	EYES	SEX
11-21-21	144	5' 5½"	LT. BLOND	BLUE	MALE

THIS CERTIFICATE is of 60 days' duration and, unless the holder hereof is otherwise notified by the Authority within such period, shall continue in effect indefinitely thereafter, unless suspended or revoked by the Authority, except that it shall immediately expire (1) at the end of each TWELVE months' period after the date of issuance hereof if the holder of this certificate fails to secure an endorsement by an authorized Inspector of the Authority within the last 45 days of each such period, or (2) at any time an authorized Inspector of the Authority shall refuse to endorse this certificate after inspection or examination.

Endorsement Refused:
Date:

Inspector, Civil Aeronautics Authority

Date of Issuance MAY 29, 1941
By direction of the Authority: Kenneth O. Wright
Signature: KENNETH O. WRIGHT
Title: ASST. AERO. INSPECTOR

This certificate is not valid unless there is attached hereto the appropriate Rating Record bearing the above number. Any alteration of this certificate is punishable by a fine of not exceeding $1,000 or imprisonment not exceeding three years, or both.

(Over)

Signature of Holder: Lewis Arthur Patteson

Photo 57: 1941 - Airman Certificate.

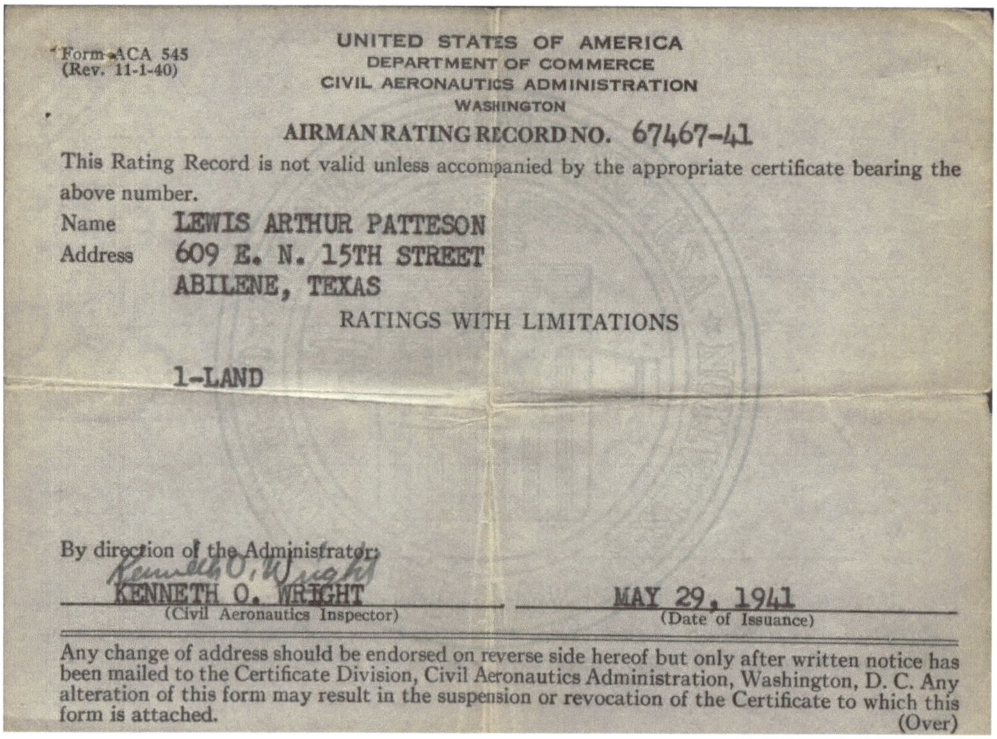

Form ACA 545
(Rev. 11-1-40)

UNITED STATES OF AMERICA
DEPARTMENT OF COMMERCE
CIVIL AERONAUTICS ADMINISTRATION
WASHINGTON

AIRMAN RATING RECORD NO. 67467-41

This Rating Record is not valid unless accompanied by the appropriate certificate bearing the above number.

Name LEWIS ARTHUR PATTESON
Address 609 E. N. 15TH STREET
ABILENE, TEXAS

RATINGS WITH LIMITATIONS

1-LAND

By direction of the Administrator: Kenneth O. Wright
KENNETH O. WRIGHT
(Civil Aeronautics Inspector)

MAY 29, 1941
(Date of Issuance)

Any change of address should be endorsed on reverse side hereof but only after written notice has been mailed to the Certificate Division, Civil Aeronautics Administration, Washington, D. C. Any alteration of this form may result in the suspension or revocation of the Certificate to which this form is attached.

(Over)

Photo 58: 1941 - Airman Rating.

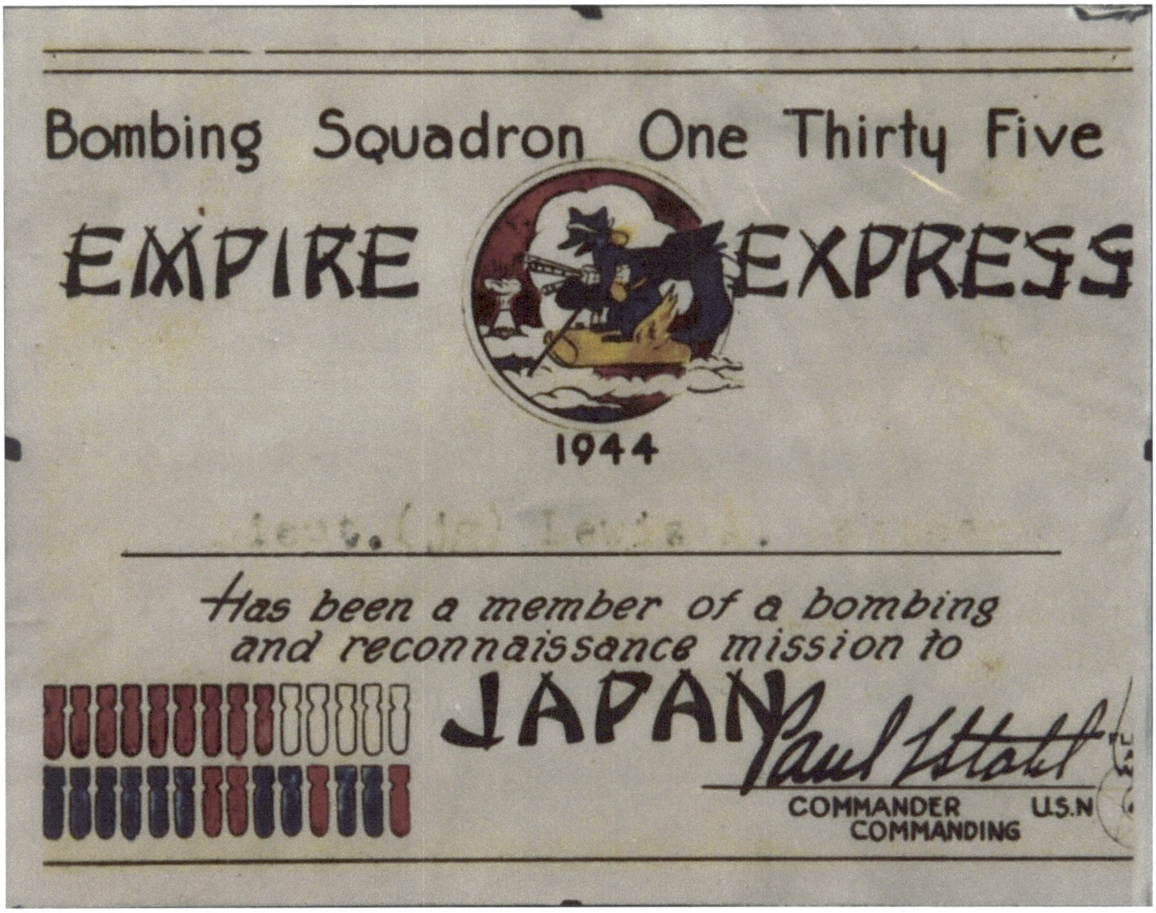

Bombing Squadron One Thirty Five

EMPIRE EXPRESS

1944

Lieut.(jg) Lewis A.

Has been a member of a bombing and reconnaissance mission to

JAPAN

COMMANDER U.S.N
COMMANDING

Photo 59: 1944 - Empire Express Card.

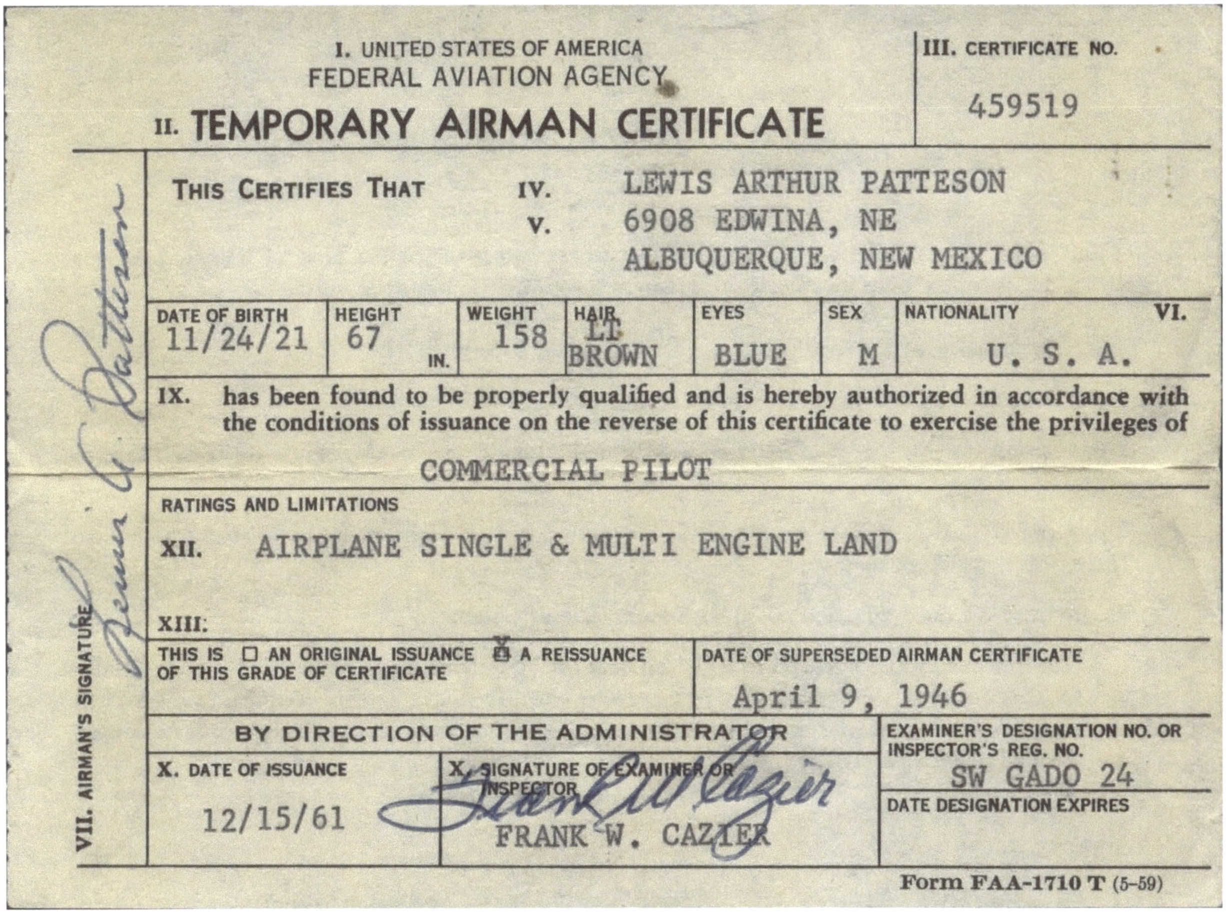

I. UNITED STATES OF AMERICA
FEDERAL AVIATION AGENCY

II. TEMPORARY AIRMAN CERTIFICATE

III. CERTIFICATE NO. 459519

THIS CERTIFIES THAT IV. LEWIS ARTHUR PATTESON
V. 6908 EDWINA, NE
ALBUQUERQUE, NEW MEXICO

DATE OF BIRTH	HEIGHT	WEIGHT	HAIR	EYES	SEX	NATIONALITY VI.
11/24/21	67 IN.	158	LT BROWN	BLUE	M	U. S. A.

IX. has been found to be properly qualified and is hereby authorized in accordance with the conditions of issuance on the reverse of this certificate to exercise the privileges of

COMMERCIAL PILOT

RATINGS AND LIMITATIONS

XII. AIRPLANE SINGLE & MULTI ENGINE LAND

XIII:

THIS IS ☐ AN ORIGINAL ISSUANCE ☒ A REISSUANCE OF THIS GRADE OF CERTIFICATE

DATE OF SUPERSEDED AIRMAN CERTIFICATE: April 9, 1946

BY DIRECTION OF THE ADMINISTRATOR

X. DATE OF ISSUANCE: 12/15/61

X. SIGNATURE OF EXAMINER OR INSPECTOR: FRANK W. CAZIER

EXAMINER'S DESIGNATION NO. OR INSPECTOR'S REG. NO.: SW GADO 24

DATE DESIGNATION EXPIRES

VII. AIRMAN'S SIGNATURE

Form FAA-1710 T (5-59)

Photo 60: 1961 - Airman Certificate.

Photo 61: WWII United States Naval Aviator Lewis Arthur "Pat" Patteson.

Photo 62: This shows the act of trying to load a Mark 13 torpedo into one of the Venturas. The "Widow Maker" of Adak. This torpedo killed more Allied servicemen than Japs. It simply was not capable of being dropped from an aircraft at any speed or altitude.

Photo 63: Plane Captain Ed Hillard #18 crew in 1944.

Photo 64: Pilot Kay Walsh #18 crew in 1944. "Pip Squeak" was a "Hedron" plane that was recovered and used for PATWING-4 activities. Not our VPB plane.

Photo 65: L. A. "Pat" Patteson #18 crew in 1944.

Photo 66: Beasly with his Brownie camera.

Photo 67: 1944 - Attu - Our Executive Officer of VPB-135, M.A. Mason. Became Commanding Officer of 3rd Squadron to Attu in 1945 with PV-2s.

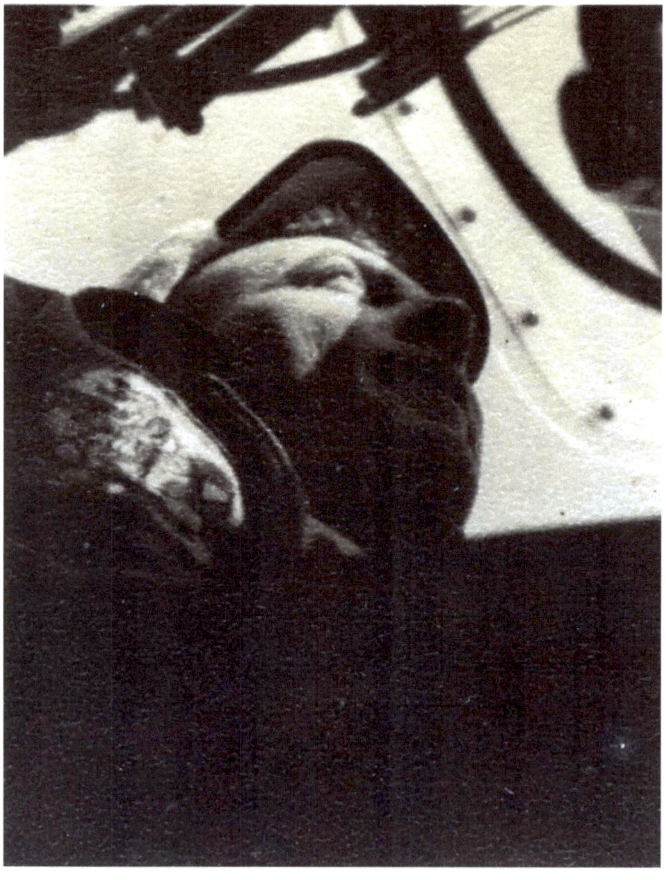

Photo 68: Turret Gunner Bedenik.

Photo 69: Attu reading room at the Photo Lab.

Photo 70: VPB-135, Attu - Mission takeoffs are only minutes away. Note: Cloud layer is less than 500 feet.

Photo 71: 1943 - Aboard Coast Guard Frigate. Alaska Inside Passage has scenery that is unimaginable.

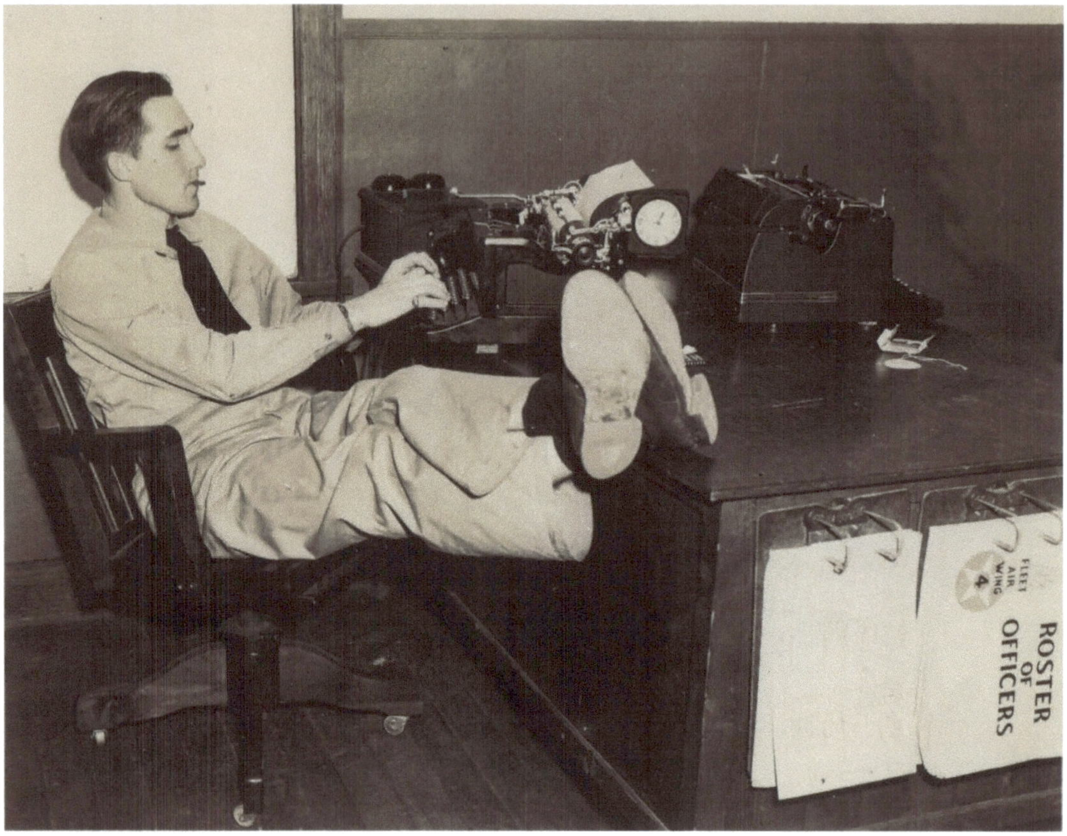

Photo 72: 1944 - Pilot Jim Rumford

Photo 73: The skipper of Coast Guard Frigate on the bridge navigating the Alaskan Inside Passage.

Photo 74: 1942 - Adak - Joe E. Brown visited the troops and sailors with a 1-man show. Lost his son in the war, but continued to go to any area, especially to small, unpopular spots like the Aleutians.

Photo 75: Cow barn on Gheres Point with attendant and chickens. A lady cow arrives on Attu. (See story).

Photo 76: 1944 - Hedron mechanics bug hunting on the Attu tarmac.

Photo 77: 1945 - Attu - Bedenik VPB-135 crew AOM 2/C

Photo 78: 1945 Attu Refueling PV-2.

Photo 79: 1943 Short Snortr Silver Certificate. Signed by Patwing-4 Commodore L.E. Gehres who gave a welcome for our arrival on Attu.

Photo 80: 6-1-43 Reverse side Short Snorter Certificate. Silver Certificates were taken out of circulation for Federal Reserve notes. We're off the gold standard.

Photo 81: Snoqualmie Pass begineers Grimstead, GF, Nichlas, Me

Photo 82: Snoqualmie Pass begineers "Nick"

Photo 83: Snoqualmie Pass begineers "Grinny"

Photo 84: Snoqualmie Pass begineers "Pete"

Photo 85: Snoqualmie Pass begineers "Pat"

Photo 86: 1942 - Ski Resort Snoqualmie Pass, WA

Photo 87: Aboard Coast Guard Frigate enroute from Kodiak to Seattle.

Photo 88: On the Canadian border.

Photo 89: 1945 - Sailboating on Lake Washington near Seattle.

Photo 90: I hold my Argus C3 camera, a Patwing 4 censured item.

Photo 91: 1944 - 1945 - Attu - Ens. Tom Macleod sneaks his censured 35 mm camera, film, developing tank and chemicals through inspections to get many secret photos.

Photo 92: 1944 - Attu - Lt. Pirtle, an ex-instructor and excellent instrument flier.

Photo 93: 1944 - Massacre Bay, Attu - Lt.(jg) Jack English. Over 20 missions, decorated and very good pilot. His name, "Jack," was also his order when he had a drink.

Photo 94: 1944 - Ens. Robillard, a funny story teller.

Photo 95: 1944 - Skipper of Coast Guard frigate bound for Seattle with several VPB-135 veterans off to form new PV-2 squadron.

Photo 96: 1943 - Lt.(jg) Tom Macleod on ferry boat to Mukilteo and Seattle.

Photo 97: 1944 - Attu - Hut photo of Kay Walsh with pictures of his GF in Kansas City. He didn't drink, smoke and only read cartoon books.

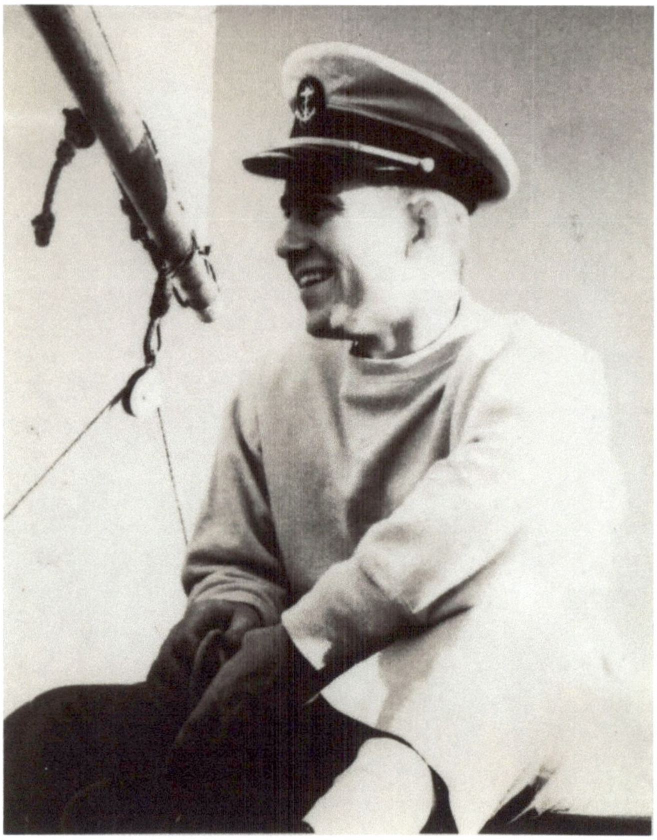

Photo 98: 1943 - Brother Morris at the helm of "our" sailboat on Galveston Bay. He was a Navy Pre-Med student.

Photo 99: 1944 - Attu - Dick Rice soaks up some sunshine outside hut #30 while re-reading a letter from his Susie.

Photo 100: 1944 - Attu - Claude Britt, our enlisted navigator used for night missions as a backup for the regular navigator, Ens. Rice.

Photo 101: 1943 - Whidbey Island - Chuck Mobus insists we have our own Christmas tree.

Photo 102: 1945 - I stroll on Attu with required metal hat. A clear day for a stroll.

Photo 103: 1945 - Attu - John Biros borrows my radio for music for his hut mates.

Photo 104: Side ride for the Oldsmobile at Whidbey Island at redwood stump.

Photo 105: 1945 - Attu - John Biros pulling props before starting the cold engine.

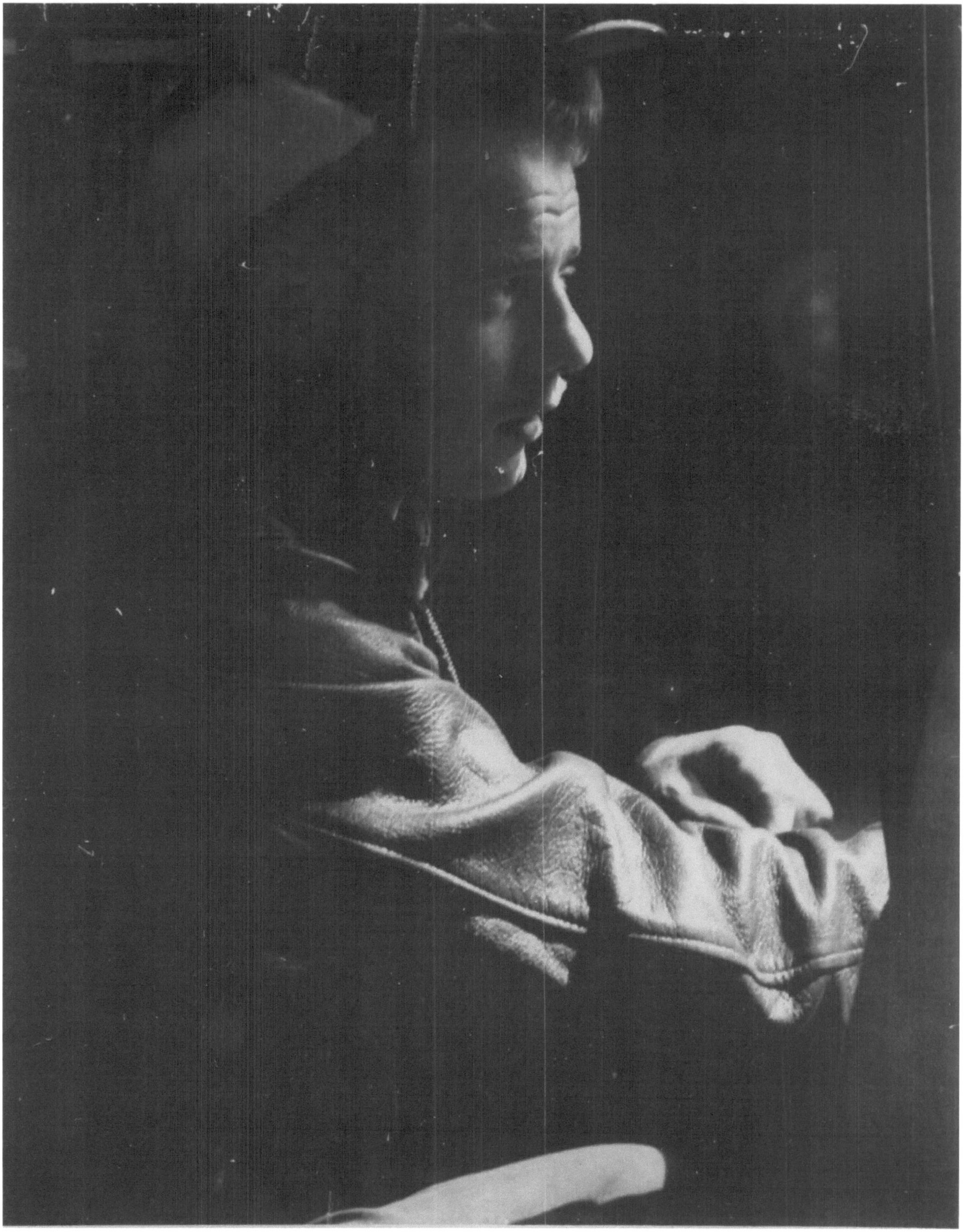

Photo 106: 1945 VPB-135 Radioman Fader.

Photo 107: 1944 - Ens. Dick Rice will dream of his Susie waiting at Minneapolis.

Photo 108: 1945 - Attu - Dick Rice at Jap dugout. One of many tunnels and other hidden storages used by the Japs during the "Battle of Attu".

Photo 109: 1944 - Attu - Jap cave at battleground. We found lots of mortar powder marked and bagged fpr use. Different measures for different ranges.

Photo 110: Snowcapped peaks block take offs and landings to the north and west runways. Gehres Point where pilots and staff live forms eastern border at Massacre Bay. "Big" Clark's crash was in center of Massacre Bay. All perished. The plane was carrying magnesium flash bombs for night photo reconnaisance.

Photo 111: 1944 - Attu - Control Tower. Massacre Bay and Gehres Point are shown to the left. The Control Tower had little authority, we landed when we damn well wanted when returning from our mission.

Photo 112: My home away from home for many months. Looking north is a beautiful scene of a snowcapped 4,000 foot hill. It was warm and heated by fuel oil. Slept six.

Photo 113: Someone "snagged" a salmon. No fishing technique required. Just cast and reel in, you're sure to snag a "Humpy."

Photo 114: 1944 - Attu - Snagging salmon. We had all we could eat.

Photo 115: 1944 - Atty - Jim Rumford and Jim Peterson relaxing with picture books at Attu photo lab.

Photo 116: 1945 - Attu - PV-2 crew #9. VPB-135. Harry Felker, Co-Pilot, self. Navigator Don Stobbs absent in hospital for appendicitis. Bottom: Sid Fader, AMR2C, John Biros, AMM2C, F. J. Bedenik, AOM3C.

Photo 117: 1945 - Attu - L-R: Sid Fader, ARM2C, John Biros, AMM2C, F. J. Bedenik, AOM2C.

Photo 118: 1945 Attu VPB-135 crew; Fader, Biros, Bedenik

Photo 119: 1944 - Attu - VPB-135 PV-2 crew.

Photo 120: 1944 - Attu - VPB-135 PV-2 crew.

Photo 121: Rumford crew at Attu.

Photo 122: 1944 - Attu - Decoration Day in dress uniforms - VPB-135, Crew #18.

Photo 123: 1944 - Attu - VPB-135 pilots of Section "A". L-R: Gardener, Stepter, Rumford. PV-1s are parked in revetments waiting for fog to lift enough to allow their takeoff.

Photo 124: 1943 - Adak - Consolidated PBY-5A Flying Boat and Lockheed PV-1 Venturas.

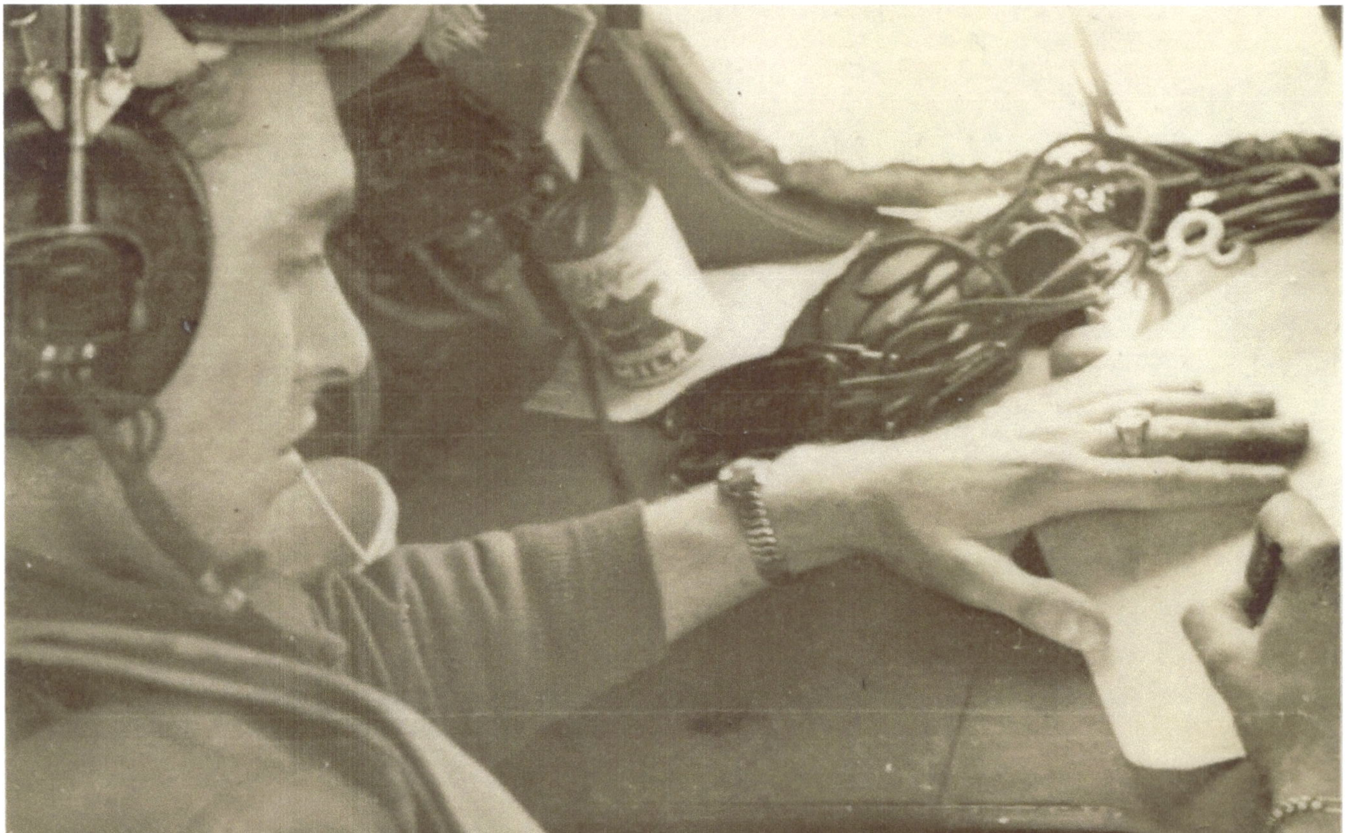

Photo 125: 1944 - L. F. Cach, ARM2C send coded messages to base Attu on return from mission. WE could send coded information when returning with results of mission and other info such as WX but cannot use radio when en route to a target.

Photo 126: My co-pilot Ensign Kay Walsh. Photo taken on mission to Kurils.

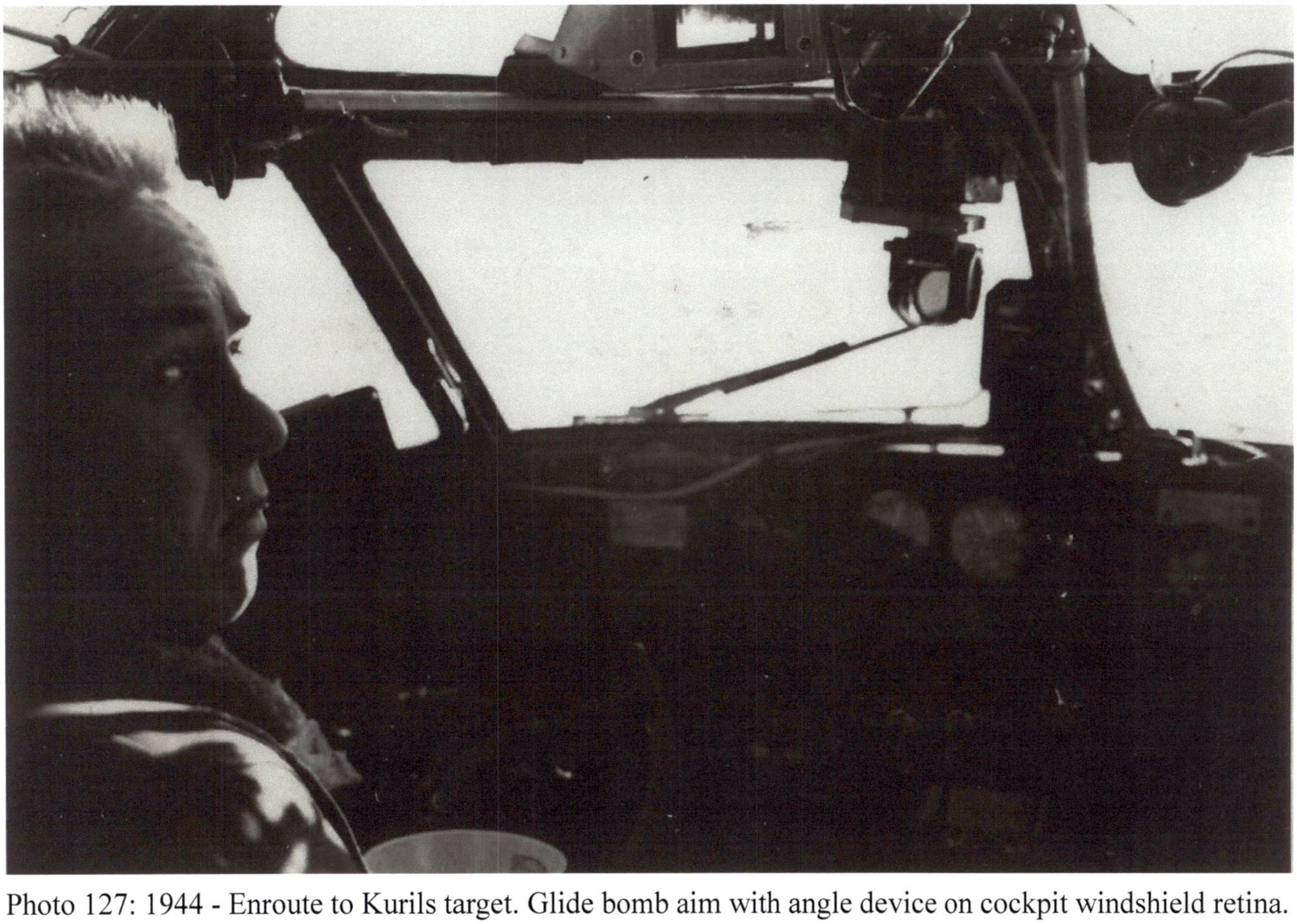

Photo 127: 1944 - Enroute to Kurils target. Glide bomb aim with angle device on cockpit windshield retina.

Photo 128: 1944 - Ens. Kay Walsh of VPB-135. Navy "co-pilot" designation is PPIP for Patrol Plane First Pilot.

Photo 129: 1945 - Ens. Harry Felker at Attu. My co-pilot. Fresh and eager from months of training in the PV-2 Harpoon.

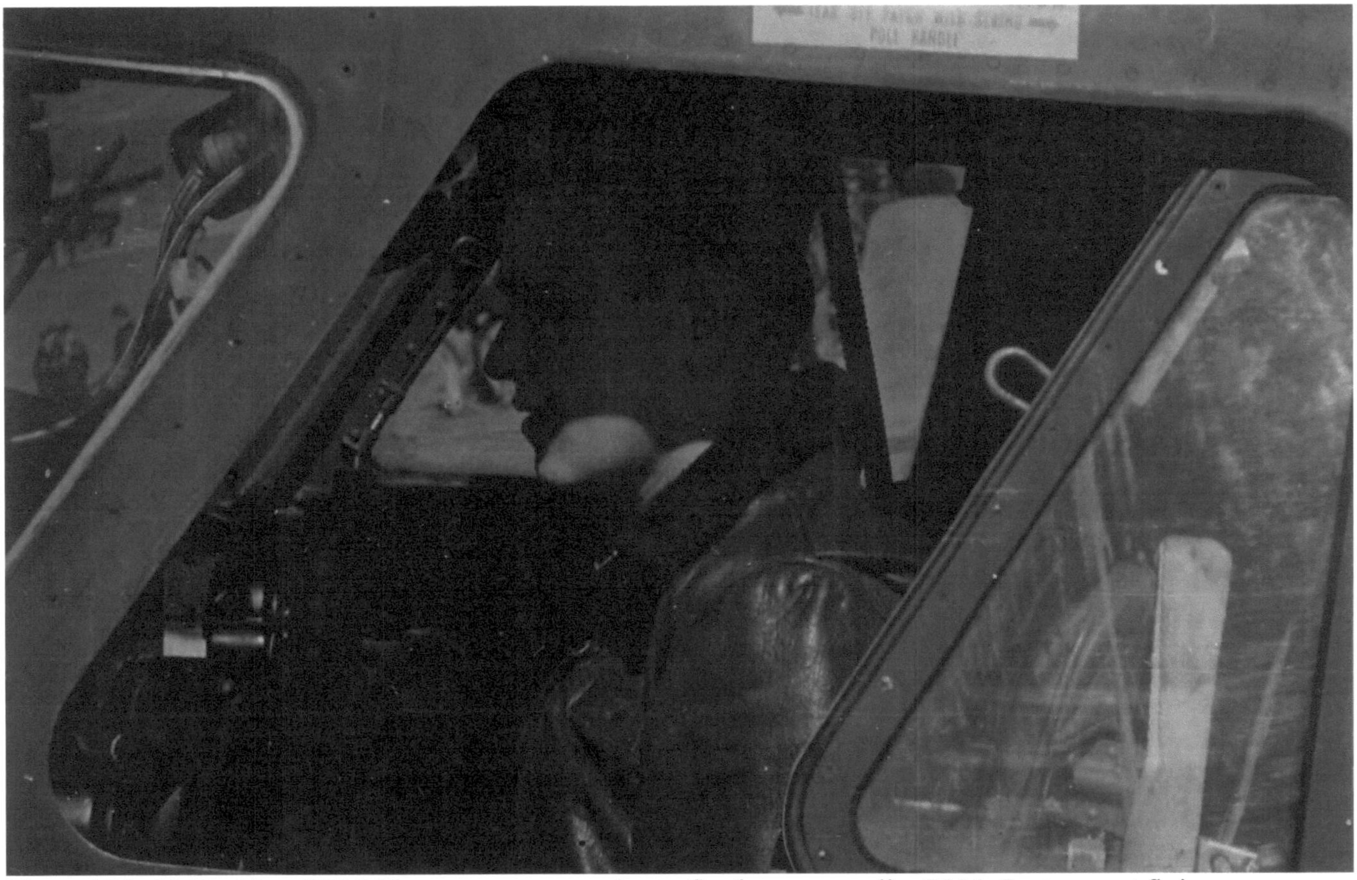

Photo 130: 1945 - New PV-2 Harpoons were easier to fly than our earlier PV-1. Instrument flying was more stable with less wandering and constant attention (drifting).

Photo 131: 1944 - Oxygen use was needed when flying above 10,000. Its use was available for entire crew. The mask was uncomfortable but definitely made a difference.

Photo 132: 1945 - Attu - Me in cockpit of Lockheed PV-2 Ventura.

Photo 133: 1944 - Refueling one of the "Cats." PBYs were in service in the Aleutians for patrol operations for entire duration of WWII. The PBY-5As used Naval Air Station Attu (Casco Cove) not the Alexai Point Army Airfield operated by the United States Army Air Forces (USAAF) also located on Attu.

Photo 134: 1944 - A group of expectant servicemen gathered to spot our returning planes back from a mission to the Kurils. Photo is looking south toward Massacre Bay.

Photo 135: PV-2 Harpoon waits in winter snow at Naval Air Station Attu.

Photo 136: 1944 - Attu - P-38 of 11th Air Force collapsed nose gear on landing at Alexai landing strip. No injuries.

Photo 137: 1944 - Amchitka - PV-1 belly landing when landing gear fails. VPB-136 squadron aircraft. No injuries.

Photo 138: Another P-38 crash landing at Alexai Point, Attu.

Photo 139: 1945 - L-R: Don Stobbs, Navigator, self, Harry Felker, co-pilot, training at Ault Field, Whidbey Island. All but one eager to return to the Aleutians.

Photo 140: 1945 - Attu - Playful Lt. Jim Rumford (3rd left) always having fun especially with the new Ensigns, Johnny Vaughter, Don Stobbs and Barfield.

Photo 141: 1944 - Attu - Wintertime brings snow and blow. Most outside activity is completely stopped.

Photo 142: 1944 - Attu - My enlisted crew in dress blues have all received an Air Medal from Adm. Frank Jack Fletcher (COMPAC). L-R: Britt, Hillard, Jacobsen and Cach.

Photo 143: 1944 - Attu - Hut mates of Quonset #30. All in our "jammies" and ready for bed. The electricity goes off at 2200 hours.

Photo 144: 1944 - Attu - The Navy dentists celebrate birth of son, Donald Melvin. Wife's photo on the wall. The father sits 3rd from left.

Photo 145: 1944 - Me and Lt.(jg) Holloway play acey-deucy (backgammon) on our "off day" schedule.

Photo 146: 1944 - My navigator, Ens. Dick Rice chats with a war correspondent. The "WC" was always looking for a story or ride with someone to a raid. He rode with us once. Turned back for WX.

Photo 147: 1944 - Attu, AK - Co-pilot Tom Macleod for Lt.(jg) Chuck Mobus, pilot. Lived in nearby hut on Gehres Point near the freshly built cow barn.

Photo 148: 1944 - Attu - Kay Walsh, J. Miller playing cards in our Hut 30.

Photo 149: L-R: Ens. Ray Wesley, Lt. Bill Clapham, Lt. Jim Rumford.

Photo 150: 1944 - All concerned to see when we will get back to home base, Attu. L-R: Navigator Rice, Jacobsen, turret gunner, Ed Hillard, plane mechanic.

Photo 151: 1945 - Aboard Coast Guard Frigate enroute from Kodiak to Seattle.

Photo 152: 1942 - Dutch Harbor, AK was bombed by Japan about 6 months after Pearl Harbor raid. The Zeros caused considerable damage but few casualties. (rare photo)

Photo 153: Many fishing boats near Ketchikan.

Photo 154: View from Coast Guard Frigate Inside Passage headed to Seattle.

Photo 155: 1944 - Attu - M.A. 'Butch' Mason, Executive Officer, Paul L. 'Skipper' Stahl, Commanding Offficer. Our squadron's capable and respected commanders. He was, we thought, a great skipper. He was pretty stern but he was very easy to get along with. Stahl had great respect for 'reserve' officers, more probably than some of his Annapolis grads.

Photo 156: 1944 - Amchitka - 11th AF at Amchitka airfield with construction still in process. B-25s line up for takeoff to bomb Kiska.

Photo 157: 1944 - Attu - Home Base Attu. Returning from strike to Paramushiru. A welcome sight.

Photo 158: Fierce Willawah was destroyed PBY-5A.

Photo 159: Alexai Point, Attu - P-38 belly landing due to landing gear problem. 11th AF Fighter Squadron. Pilot was okay. Plane parts were salvaged.

Photo 160: April, 1943 - We reach Adak, AK and have first operational flights, sector searches.

Photo 161: 1943 - Two by two we head to the Aleutians. Our first stop was Ketchikan, AK.

Photo 162: Occasionally dress blues were required at Ault Field, Whidbey Island, WA. L-R: Bill Clapham, People of Color, Pilot Kay Walsh, my co-pilot, Ray Wesley, co-pilot, me, Tom Macleod, Chuck's co-pilot, Chas. Mobus, PPC. Squadron VPB-135, 1944. This occasional "dress blue" night as an orchestral dance night with the "Tommy Dorsey and his Orchestra" big band.

Photo 163: 1944 - BOQ Bachelors at Whidbey Island. L-R: Ray Wesley, Tom Macleod, Chas. Mobus, Kay Walsh and LAP. Is Seattle ready for this?

Photo 164: 1943 - Our Oldsmobile to take us to Seattle. The Whidbey Island BOQ in background at Oak Harbor, WA. It ran great on 100 octane Av. Gas. No more hitchhiking.

Photo 165: Dec. 1942 - The luxurious red brick BOQ at Naval Air Base, Seattle, WA. I was quartered here for a short while before being sent to the new NAS at Whidbey Island. This was Navy living at its best. (Adjoining 9-hole golf course and gymnasium included).

Photo 166: Mulketio Ferries cross between Seattle and Whidbey Island - the southern passage from Ault Field to Seattle.

Photo 167: Ault Field, Whidbey Island as seen from my BOQ window. Former land owner's home still in use.

Photo 168: 1944 - Whidbey Island - BOQ Bachelors: Ens. Ray Wesley, Ens. Tom Macleod, Lt. (jg) L. A. Patteson, Lt. (jg) Chas. Mobus, Ens. Kay Walsh, Lt. (jg) Bill Clapham (photog.) [colorized]

Photo 169: PV-1s head to Aleutians.

Photo 170: 1942 - PBY-5A with early identification, single star. On patrol somewhere in the Aleutians.

Photo 171: Early PBYs on patrol in Aleutians. Note radar antenna under wings.

Photo 172: 1944 - One of my training flights over Puget Sound near Victoria, BC.

Photo 173: Early OS2U-1 "Kingfisher" scout plane probably based near Seattle. Used for short inshore patrols.

Photo 174: PV-1 (no Navy serial number) used by USN and VPB squadrons.

Photo 175: Vega (Lockheed) PV-1 Ventura (Early photo).

Photo 176: Consolidated PBY-5A Catalina. The 5A had wheels that could make it amphibious, landing on land or water.

Photo 177: Famous photo formation of PBYs near mountains and glaciers (Melispina Glacier south of Kodiak).

Photo 178: VP-41 PBY-5A near Aleutians.

Photo 179: Ens. Schulte and sister at home in St. Louis, on leave. Pete was killed in crash after takeoff from Alameda NAS in 1942, first casualty of new squadron VPB-135.

Photo 180: Ketchikan docks.

Photo 181: Ketchikan docks at low tide. High tide rose 10 feet.

Photo 182: 1944 - Coast Guard frigate somewhere in Inland Passage. I was aboard returning from Kodiak to Seattle in 1943 to train for the 2nd deployment of VPB-135.

Photo 183: Reliable Pratt & Whitney R-2800 engines take us near many volcanic peaks, a few still smoking and active. If our purpose was scenic beauty we would be enjoying one of the most majestic views on earth.

Photo 184: 1943 - Jap hangar, roads and runway near Kashiwabara, Kurile Islands.

Photo 185: Kiska Grounded Jap freighter at Kiska.

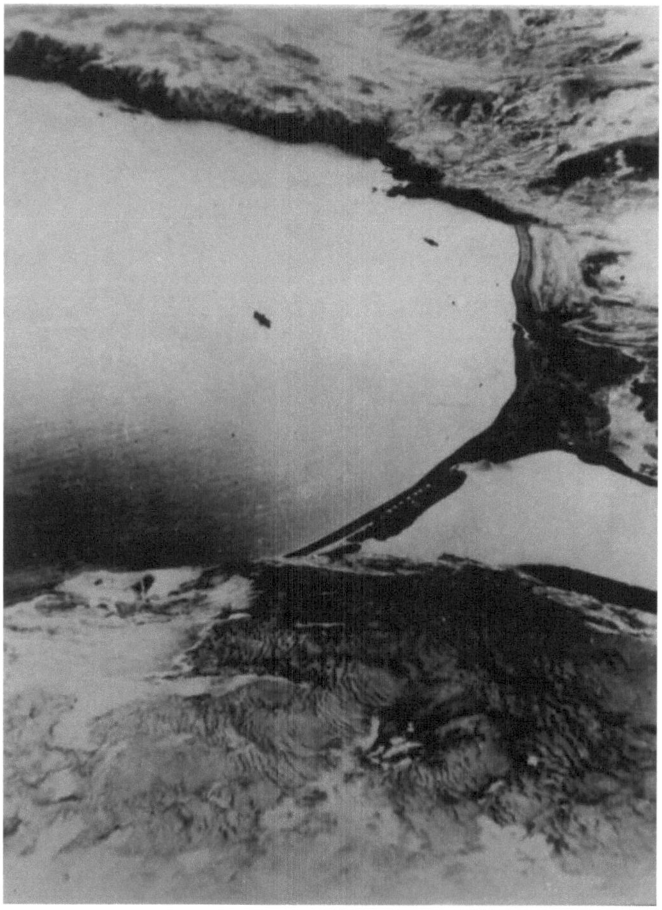

Photo 186: Grounded Jap freighters on Amchitka.

Photo 187: Jap freighter aground at Kiska.

Photo 189: 1944 - PV-1s of VPB-135 leaving revetments on Attu for mission to Kurile Islands.

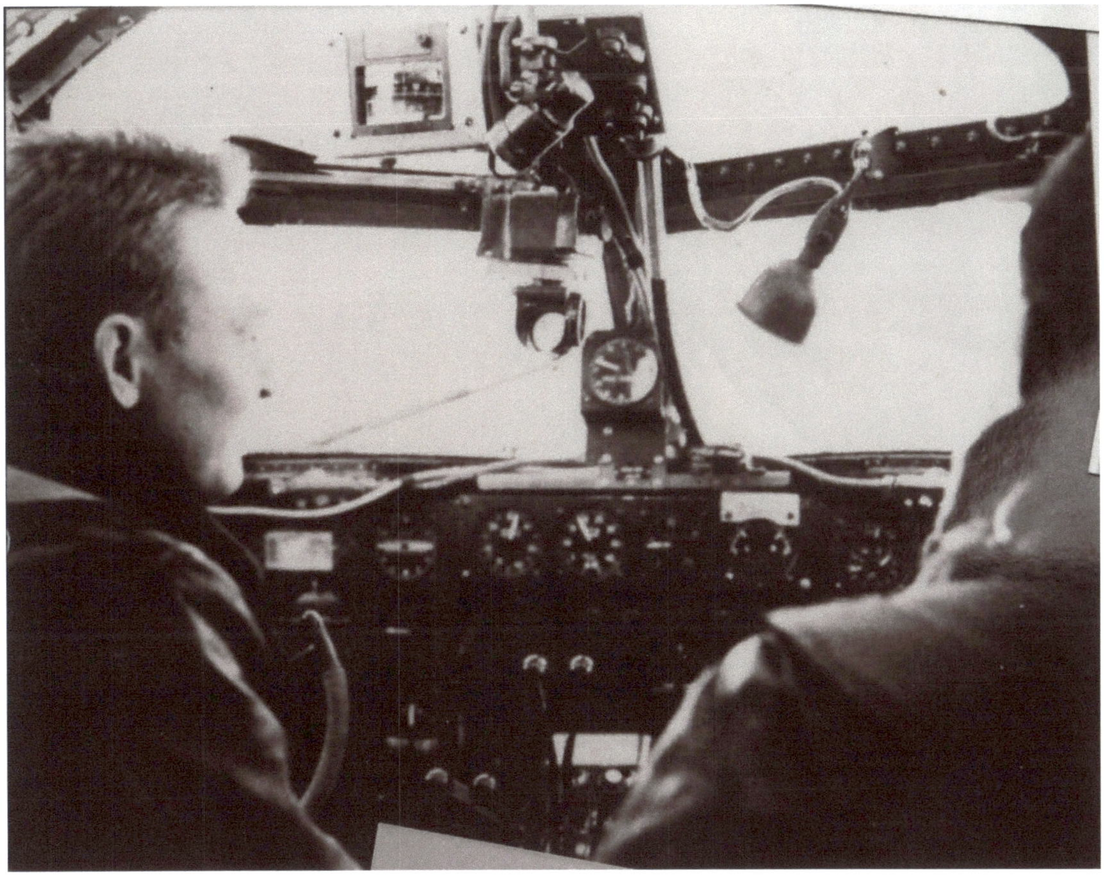

Photo 188: 1944 - Crew #18 enroute to Jap Kurile Islands. Cruising: 1900 RPM and 21 inches of Hg. On autopilot if weather is smooth.

Photo 190: 1944 - Decoration Day Ceremonies by Comm. Leslie E. Geheres. My first Distinguished Flying Cross (DFC).

Photo 191: Adm. COMNORPAC Frank Jack Fletcher - Attu.

Photo 192: Pat's leather naval aviator "soft patch" worn on his flight jacket. The naval aviator wings insignia is a warfare qualification of the United States military that is awarded to those aviators of the United States Navy, Marine Corps, and Coast Guard who have qualified as naval aviators. The badge was designed by John H. Towers c. 1917 and consists of a single fouled anchor, surmounted by a shield with 13 stripes, centered on a pair of wings.

Photo 193: Pat has 2 Distinguished Flying Crosses, 2 Air Medals and an Asiatic–Pacific Campaign Medal.

In reply address not the signer of this letter, but Bureau of Naval Personnel, Navy Department, Washington, D. C.
Refer to No.

Pers-328-101-RDQ
130094

NAVY DEPARTMENT
BUREAU OF NAVAL PERSONNEL
WASHINGTON 25, D. C.

FOR VICTORY BUY UNITED STATES WAR BONDS AND STAMPS

26 September 1946

To : Lieutenant Lewis A. Patterson, USNR,
Smiley, Texas.

Via:

Subj: Transmittal of permanent citation(s) for the Distinguished Flying Cross.

1. There is forwarded herewith the permanent citation(s) for subject decoration(s) awarded you for service during the war.

2. The records of the Chief of Naval Personnel indicate that you have been presented with the subject decoration(s).

By direction of the Chief of Naval Personnel.

L. C. Thompson

L. C. THOMPSON,
Lt. Comdr., USNR,
Medals and Awards.

Encl:
A. Permanent Citation(s) (1)

C 65769

Photo 194: Transmittal of permanent citation for the Distinguished Flying Cross 25SEP1946.

1st AM

The President of the United States takes pleasure in presenting the AIR MEDAL to

LIEUTENANT LEWIS ARTHUR PATTESON
UNITED STATES NAVAL RESERVE

for service as set forth in the following

CITATION:

"For meritorious achievement in aerial flight as Pilot of a Medium Landplane Bomber in action against enemy Japanese forces in the North Pacific Area on June 12, 1944. Voluntarily participating in the extremely hazardous task of reconnoitering the West Coast of Paramushiru Island, Lieutenant (then Lieutenant, Junior Grade,) Patteson sighted the long sought Japanese airfield at Daigo Saki and installations at Kakumabetsu Wan through a break in the fog and launched a low altitude bombing and strafing attack, scoring accurate hits on the target area, and despite cross-fire of enemy guns, obtaining photographs which were invaluable in plotting these installations for future operations. His airmanship and courage were in keeping with the highest traditions of the United States Naval Service."

For the President,

James Forrestal
Secretary of the Navy

Photo 195: Citation for 1st Air Medal for "Patteson sighted the long sought Japanese airfield at Daigo Saki... through a break in the fog and launched a low altitude bombing and strafing attack...."

THE SECRETARY OF THE NAVY
WASHINGTON

The President of the United States takes pleasure in presenting the DISTINGUISHED FLYING CROSS to

LIEUTENANT LEWIS ARTHUR PATTESON
UNITED STATES NAVAL RESERVE

for service as set forth in the following

CITATION:

"For heroism and extraordinary achievement in aerial flight as Commander of a Ventura Medium Bomber in Bombing Squadron ONE THIRTY FIVE attached to Fleet Air Wing FOUR, during action against enemy Japanese forces in the Islands of Shimushu and Paramushiru, June 14, 1944. Volunteering to accompany a daylight strike on Miyoshino Airfield in a specially equipped photographic plane, Lieutenant (then Lieutenant, Junior Grade,) Patteson carried out his hazardous flight without fighter escort or cloud cover. Although intercepted by enemy fighters and subjected to antiaircraft fire, he steadfastly held his course and obtained excellent photographic coverage of the assigned area. Finding the southern portion of Paramushiru open, he boldly made a second run over the Kurabu Zaki area and, despite renewed contact with intercepting fighters, successfully completed his mission. Lieutenant Patteson's skilled airmanship, gallant initiative and daring tactics reflect the highest credit upon himself and the United States Naval Service."

For the President,

James Forrestal

Secretary of the Navy

Photo 196: Citation for 1st Distinguished Flying Cross for actions June 19th 1944 over the Miyoshina and Kurabi Zaki airfields on the islands of Shimushu and Paramshiro.

In the name of the President of the United States, Commander North Pacific Force, United States Pacific Fleet, takes pleasure in awarding the GOLD STAR in lieu of a second DISTINGUISHED FLYING CROSS to

LIEUTENANT (JG) LEWIS A. PATTESON
A-V(N)
UNITED STATES NAVAL RESERVE

for service as set forth in the following

CITATION:

"For heroism and extraordinary achievement in aerial flight. On 25 June 1944, Lieutenant (jg) Lewis A. Patteson, A-V(N), U. S. Naval Reserve, as patrol plane commander of a Ventura bomber on a combined weather flight and a daylight photo-reconnaissance of the Kurile Islands, Japan, finding his designated areas overcast, proceeded to the Kakumabetsu airfield area, where he obtained complete photo-coverage of the airstrip under construction, and a total of 121 excellent vertical and oblique photographs of installations at Kakumabetsu, Kashiwabara, and the north-eastern coastline of Paramushiro. He then proceeded northward over the island of Shimushu in clear weather, despite the probability of interception by enemy fighters. Shortly thereafter he was attacked by eleven Hamp type enemy fighters. In the ensuing action, two of the Japanese planes were hit by turret gunfire and fell away smoking, one a probable kill. Due to Lieutenant Patteson's superior piloting skill and the accuracy of his gunners his own plane sustained only minor damage. His conduct throughout was in keeping with the highest traditions of the naval service."

FRANK JACK FLETCHER
Vice Admiral, U. S. Navy.

Temporary Citation

Photo 197: Temporary Citation for Gold Star for 2nd Distinguished Flying Cross for actions on photographic mission over Shimushu on 25JUN1944 including hits on two Japanese planes and a probable kill.

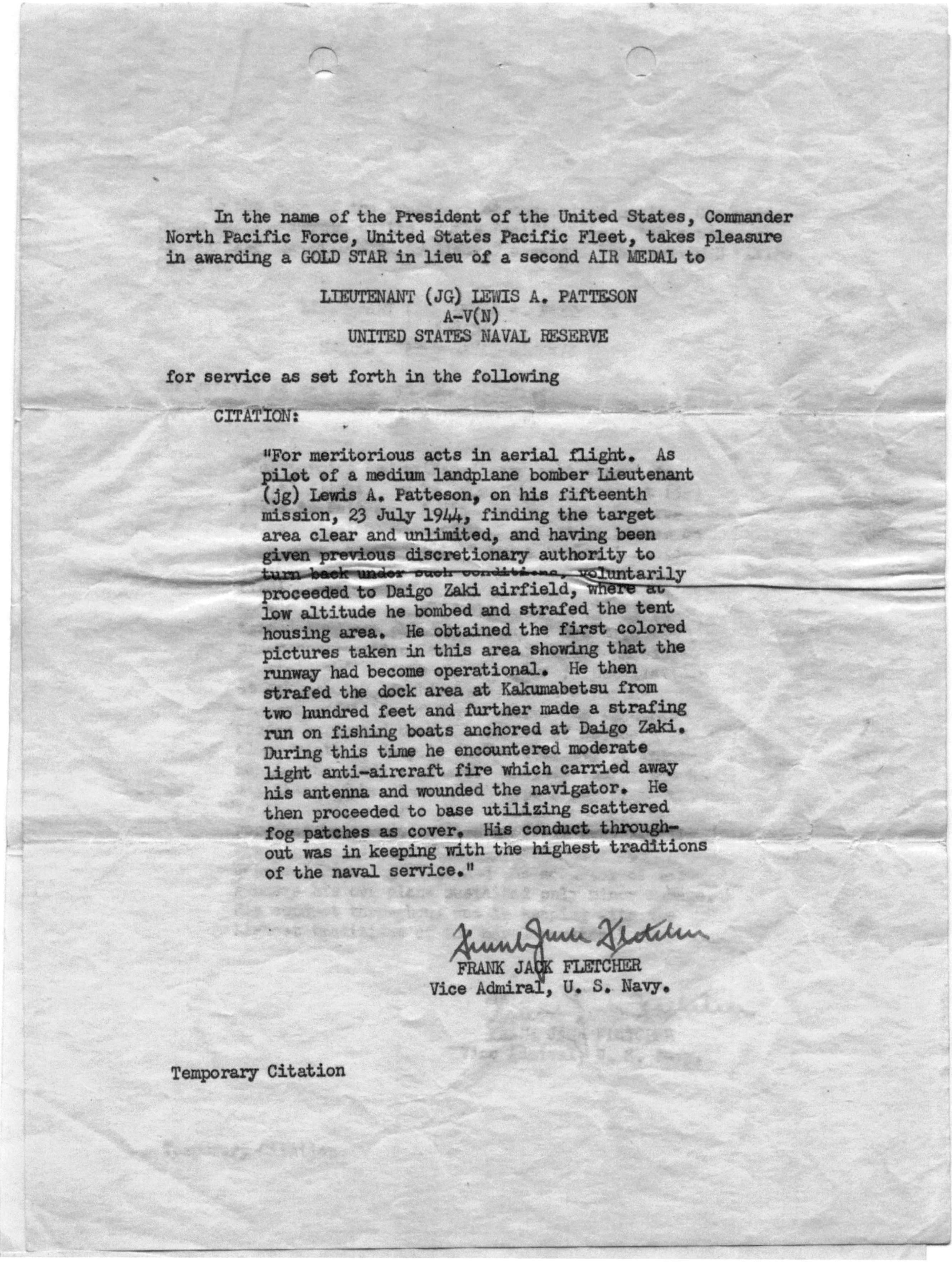

In the name of the President of the United States, Commander North Pacific Force, United States Pacific Fleet, takes pleasure in awarding a GOLD STAR in lieu of a second AIR MEDAL to

LIEUTENANT (JG) LEWIS A. PATTESON
A-V(N)
UNITED STATES NAVAL RESERVE

for service as set forth in the following

CITATION:

"For meritorious acts in aerial flight. As pilot of a medium landplane bomber Lieutenant (jg) Lewis A. Patteson, on his fifteenth mission, 23 July 1944, finding the target area clear and unlimited, and having been given previous discretionary authority to ~~turn back under such conditions,~~ voluntarily proceeded to Daigo Zaki airfield, ~~where at~~ low altitude he bombed and strafed the tent housing area. He obtained the first colored pictures taken in this area showing that the runway had become operational. He then strafed the dock area at Kakumabetsu from two hundred feet and further made a strafing run on fishing boats anchored at Daigo Zaki. During this time he encountered moderate light anti-aircraft fire which carried away his antenna and wounded the navigator. He then proceeded to base utilizing scattered fog patches as cover. His conduct throughout was in keeping with the highest traditions of the naval service."

FRANK JACK FLETCHER
Vice Admiral, U. S. Navy.

Temporary Citation

Photo 198: Temporary citation for Gold Star for 2nd Air Medal for action 23JUL44 against Daigo Zaki airfield.

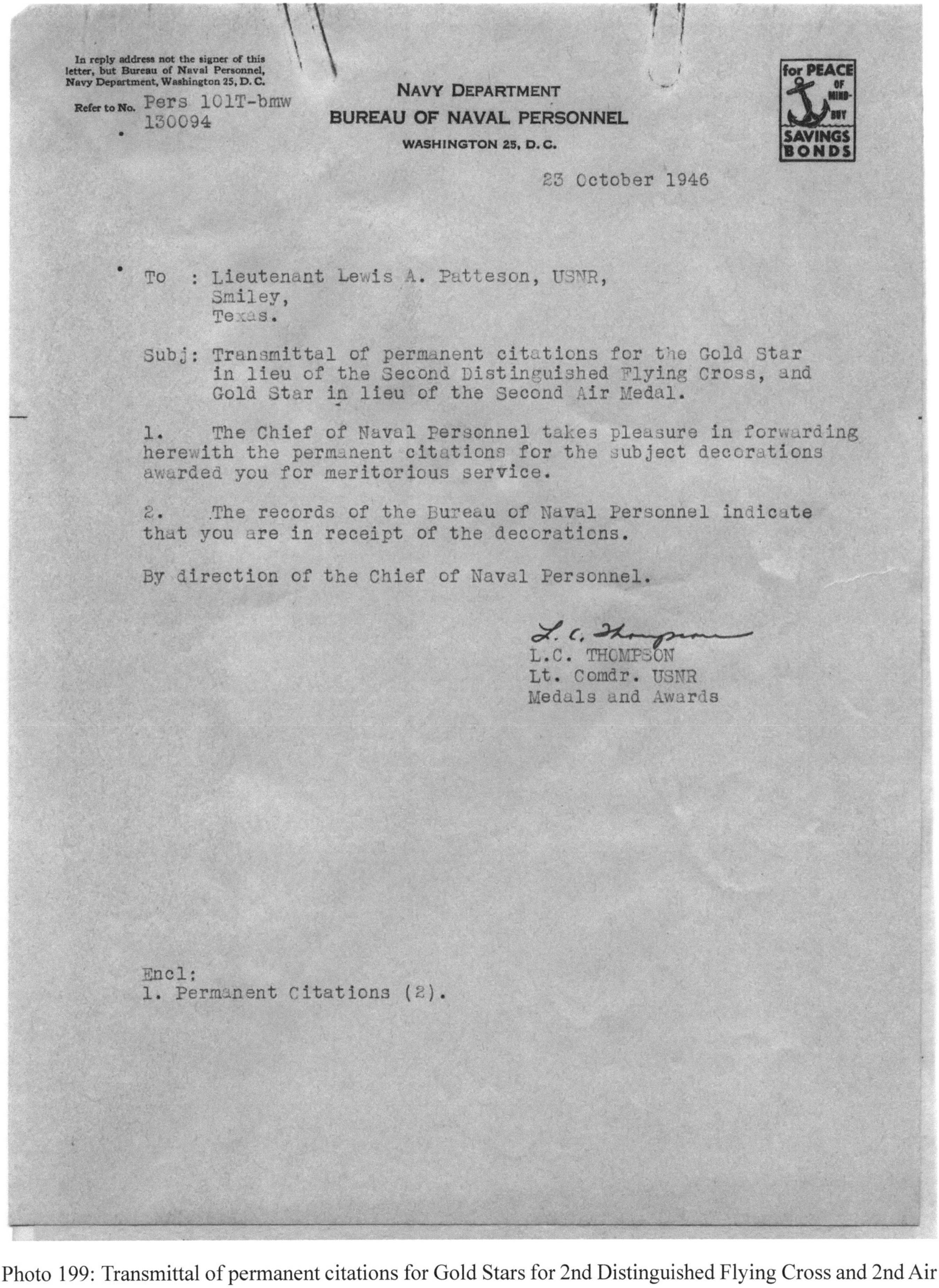

In reply address not the signer of this letter, but Bureau of Naval Personnel, Navy Department, Washington 25, D. C.

Refer to No. Pers 101T-bmw
130094

NAVY DEPARTMENT
BUREAU OF NAVAL PERSONNEL
WASHINGTON 25, D. C.

for PEACE OF MIND - BUY SAVINGS BONDS

23 October 1946

To : Lieutenant Lewis A. Patteson, USNR,
Smiley,
Texas.

Subj: Transmittal of permanent citations for the Gold Star in lieu of the Second Distinguished Flying Cross, and Gold Star in lieu of the Second Air Medal.

1. The Chief of Naval Personnel takes pleasure in forwarding herewith the permanent citations for the subject decorations awarded you for meritorious service.

2. The records of the Bureau of Naval Personnel indicate that you are in receipt of the decorations.

By direction of the Chief of Naval Personnel.

L. C. Thompson
L.C. THOMPSON
Lt. Comdr. USNR
Medals and Awards

Encl:
1. Permanent Citations (2).

Photo 199: Transmittal of permanent citations for Gold Stars for 2nd Distinguished Flying Cross and 2nd Air Medal 23OCT1946.

THE SECRETARY OF THE NAVY
WASHINGTON

The President of the United States takes pleasure in presenting the GOLD STAR in lieu of a Second Distinguished Flying Cross to

LIEUTENANT LEWIS ARTHUR PATTESON
UNITED STATES NAVAL RESERVE

for service as set forth in the following

CITATION:

"For meritorious achievement in aerial flight as Patrol Plane Commander of a Ventura Bomber, during a combined weather flight and daylight photo-reconnaissance mission in the Kurile Islands, Japan, June 25, 1944. Cool and courageous throughout this hazardous mission, Lieutenant (then Lieutenant, Junior Grade,) Patteson assisted in obtaining complete photographic coverage of the airstrip under construction and in securing a total of 121 excellent vertical and oblique photographs of installations at Kakumabetsu, Kashiwabata and the north-eastern coastline of Paramushiro. Fighting his plane with skill and daring when he was attacked by eleven Japanese fighters, he enabled his gunners to destroy or damage two of the hostile aircraft while his own plane sustained minor damage. His superior airmanship and devotion to duty reflect the highest credit upon Lieutenant Patteson and the United States Naval Service."

For the President,

James Forrestal

Secretary of the Navy

Photo 200: Citation for Gold Star for 2nd Distinguished Flying Cross for actions on daylight photo-reconnaissance mission 25Jun1944 damaging 2 Japanese fighters.

THE SECRETARY OF THE NAVY
WASHINGTON

The President of the United States takes pleasure in presenting the GOLD STAR in lieu of a Second Air Medal to

LIEUTENANT LEWIS ARTHUR PATTESON
UNITED STATES NAVAL RESERVE

for service as set forth in the following

CITATION:

"For meritorious achievement in aerial flight as Pilot of a Medium Landplane Bomber operating over the enemy Japanese-held Kuriles, on his fifteenth mission, July 23, 1944. Unable to bomb and strafe his original target, Lieutenant (then Lieutenant, Junior Grade,) Patteson voluntary proceeded to Daige Zaki where, in the face of light antiaircraft fire which took away his antenna and wounded his navigator, he bombed and strafed the tent housing area, the docks and fishing boats at Kakumabetsu and obtained the first colored pictures showing that the runway had become operational. His courage in the face of danger, his devotion to duty, and his perseverance were in keeping with the highest traditions of the United States Naval Service."

For the President,

James Forrestal

Secretary of the Navy

Photo 201: Citation for Gold Star for 2nd Distinguished Flying Cross for voluntarily flying his "Medium Landplane Bomber" over enemy territory returning the 1st color photographs showing the runway at Daigo Zaki was operational.

Photo 202: Alaskan Aviation Heritage Museum. I donated the chronometer, lower right.

Photo 203: Model of Lockheed PV-1 Ventura. Constructed by Boris Ilchenko.

Photo 204: Model of Lockheed PV-1 Ventura.

Sept 27, '95

Dear Pat: As you can see, I've got the darkroom cranked up again and I've put your collection on the top of the stack. We were out of action there for awhile due to the lack of trained personnel. Hope to have your photo copies (5x7's) to you within the month.

After our discussion in Tacoma I've decided to visit & treat the PV1 Ventura on Agattu. I think we're going to bring it in before it goes downhill any further. Hope all is well with you.

Best Regards, T.M. Spencer

Ted Spencer
AK Avia Heritage Museum
4721 Aircraft Drive
Anchorage, AK 99502-1080

Pat Patteson
9 Everett Court
Danville, CA 94526

Photo 205: Postcard from Ted Spencer about a PV-1 that made an emergency landing on Agattu Island. The plane was in squadron VPB-136 and WX was fogged in and low of fuel . Crew all waded to to land and and waited for rescue. The plane was never recovered.

Photo 206: On December 30, 1943 took off from Attu Airfield on Attu Island piloted by V. C. Austin on a search mission over sector "George 8". Returning low on fuel, this aircraft force landed with the landing gear up near a frozen lake on Agattu Island, approxiamtely 30 miles SE from their base. (Pacific Wrecks)

Photo 207: Navy PV-1 Ventura patrol aircraft of Bombing Squadron VB-139 returning to Attu was forced down due to low fuel and landed on a frozen lake, Agattu Island, Aleutian Islands, Alaska, 30 Dec 1943 The crew suffered minor injuries during the landing. (WWII Database)

Photo 208: The wreckage of this aircraft remains in situ to this day. The tail section is broken off and left outer wing and engine are broken off. (Pacific Wrecks)

Photo 209: Melissa's 40th Birthday Party with Pat. Lots of fun!.

Photo 210: Melissa's 40th Birthday Party with beaus; Alex and Lea.

Photo 211: 2015 - Lots of interesting sights to see at the show. Reno Air Show AGM "Zero," Travis AFB.

Photo 212: San Francisco 2019 Honor flight. I'm briefed on the flight with the captain before takeoff. Non-stop from San Francisco SFO to DCA.

Photo 213: I fly again in PV-2 "Harpoon."

Photo 214: 2015 - Flew to Travis AFB.

Photo 215: Starboard engine oil leak caused emergency landing at Chino. The PV-2 was scheduled to be taken aboard the USS Essex Carrier for participation in air show at Honolulu.

Photo 216: Debbie Hammond's surprise visit from the Admiral.

Photo 217: Pat with WWII Veteran.

Photo 218: Pat back in the cockpit.

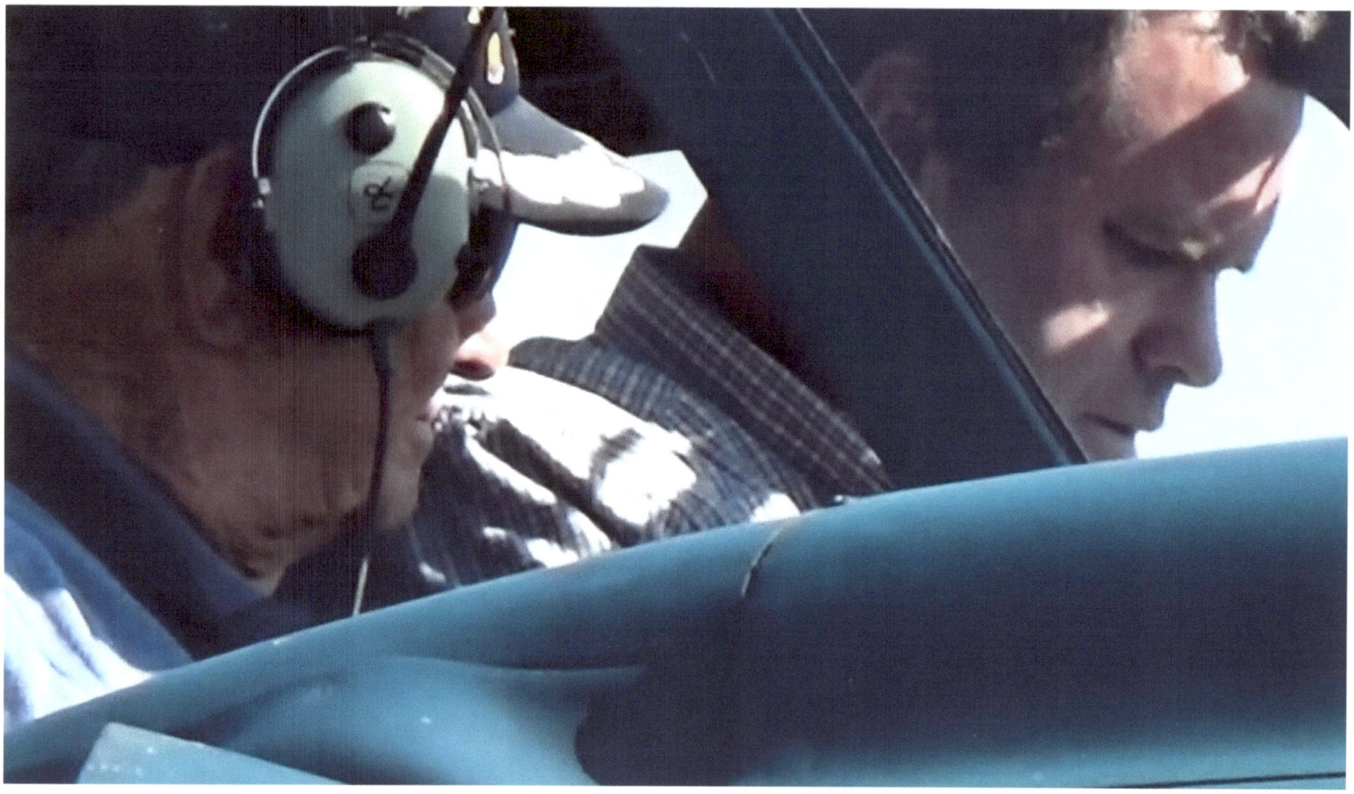

Photo 219: Pat with Taigh.

Photo 220: CONTACT! Hand-propping Taigh Ramey's restored PV-2 Harpoon.

Photo 221: Airport centerpiece for new terminal at Sanford, FL. PV-1 lifted into it's location for public view. Last Ventura for viewing in the USA.

Photo 222: The Orlando Sanford International Airport recently opened the new section of their airport terminal which includes the display of a PV-1 Ventura. This PV-1 Ventura was offered to the Sanford Airport Memorial Committee by the Museum of Naval Aviation for display in the airpark. Florida Warbirds photo. [https://floridawarbirds.com/2014/05/25/sanford-naval-air-station-ventura-restoration/]

Photo 223: 8-13-2021 San Jose 'History Park' with Zelda a "Rosie the Riveter" and a lineup of WWII Service veterans are honored.

Photo 224: We meet these 2nd graders at Stead AFB, Reno, NV. When greeted some were unable to remember their own name.

Photo 225: Oct. 2019 Reno Air Races. The 'Zero' of WWII, a flying Mitsubishi A6M.

Photo 226: PV-2 Harpoon on exhibit at Reno Air Show Oct. 2019.

Photo 227: Hands on throttles of PBY-5A from Napa.

Photo 228: 10-16-2020 Napa, CA. I sit in pilot seat of PBY-5A. Overhead controls for trim tabs, propeller and carburetor controls.

Photo 229: May, 2022, AAF visitors inspect our PV-2 Harpoon at Travis AFB.

Photo 230: D-Day Veteran Jake Larson gives thumbs up with WWII Veterans at the Third Thursday Lunch Bunch celebrating Pearl Harbor Survivor Earl J. "Chuck" Kohler's 99th Birthday Party Jan. 19, 2023.

Photo 231: Pat's WWII scrapbook is described photo by photo in the interview in the next section.

Photo 232: Pat at home recording the interview.

Early Life In Smiley, Flight Training, VPB-135 Deployment

Editor's Note: This is a 45-minute interview where Pat discusses growing up and the first part of his military career. We go through one of his binders of photographs and Pat describes each page.

Pat: I can begin, I guess, at the beginning and that was in a little village named Smiley in Texas, population of around 600 on November 24, Thanksgiving Day, my mother gave birth to twin boys. I was one of them and we grew up in this rural community, primarily farms and ranches and very easygoing but hardworking villagers. We were on one of the main roads in Texas that went from Victoria to San Antonio, Texas. We were about 90 miles south of Austin.

We both went to the school, which was not only a high school but a grammar school and at that time school graduated you in 11 grades, not 12. My brother and I made good grades but he was better at academics than I was and I was more interested in activities. I wasn't very large so I didn't make the first team on anything, but I did learn some music that my mother had taught me and she began me on the violin at something like five years old. I thought I was going to be good at it but I found out I was not very good. I was good enough to get a scholarship at a college in West Texas. The college was Abilene Christian College in Abilene, Texas.

It was there that my brother took pre-med. I was trying to get engineering courses but the college was so small that they didn't have good teachers for physics or mathematics and things that I knew I would need if I wanted to continue in aviation. At any rate, I did continue and the school eventually, because of the pressures of the war, WWII, and the activity in Europe, the schools were allowed or given government help in starting what they called a Civilian Pilot Training Program. Now, this was something I took in my sophomore year and it was the start of my beginning to fly in a real airplane. Before then I had simply fooled around with toys and watched the airplanes fly overhead. This training was of benefit to me because the war started, of course, on Pearl Harbor Day, December 7, 1941. I was at the time nearing my 20th birthday. Being 20 years old was the right age in you could join either the Air Force or the Naval Air Force. They both were building training schools and I was influenced by another schoolmate of mine, took a liking to the Navy approach to things and went down to the post office about a week after I turned 20 years old and joined the Navy. I got into a program called the V5 Program which was for training applicants to the Aviation Cadet Program.

I took two courses in that and made good grades, the best grades I ever made in college, were in this Civilian Pilot Training Course. The first was the primary, which had a lot of studies about air and of weather and power and how you generate power in an engine and other mathematical lessons, particularly in navigation. So that training I got before going into the Navy put me in good stead to go right on to be accepted by the Naval Cadet Program. And this started out first in basics up at Lou Foote Field near Dallas. I went through that without any problem because I had already learned to fly and solo in a plane. Then they transferred me to this large Naval training school in Corpus Christie. This was larger than the one at Florida, Pensacola, the old original training school for pilots.

From that point I went through the cadet program primary, in which we flew the old yellow biplane called the "Yellow Peril". That was supposed to solo you, but I had already soloed. I went through that with high marks. The next was basic in which you learned the Navy manner of flying formations and night flying and navigation at nighttime. Then you chose the type of final flying program you wanted. You could choose to be a fighter pilot, a dive bomber pilot, a float plane, or a PBY twin-engine aircraft. I selected the PBY because multi-engine was my desired goal in getting through, finishing at the Naval air station and getting my wings and becoming commissioned as an Ensign.

Photo 233: Famous photo formation of PBYs near mountains and glaciers including Melispina Glacier south of Kodiak.

Photo 234: Brother Morris in his premed attire.

My brother, my twin brother, was named Morris. He went through pre-med school in college and later on became a medical officer in the Navy. I had two younger sisters, one of still lives in a home near Gonzalez, Texas and her name is Nelda Jean Patteson. My other sister is deceased but she married from college and her husband was in the oil drilling business. They had three sons, very good young men and one of them was good in football, and the other two were good in other academics like engineering and architecture. So I correspond with two of them. One of them passed about five years ago. My youngest sister passed about one year ago. *Lea: And Larry passed a couple months after that.* Pat: Another son, passed about two years later.

My father was the son of a rancher who was an early Texas settler and owner of ranch land, quite a tract of land where they raised cattle and sheep and did other farming requirements for feeding the family. Dad was one of 11 children and he enlisted in WWI and went to France, served there, and came back from France as a Second Lieutenant. With his discharge money and some he had saved, he built and started a garage in the little town of Smiley, Texas, where I was born. That was during the Ford Model T days and the Ford Model A days. My brother and I helped him with that. We knew how to grease the cars, we knew how to check them out and certainly we knew how to fill them with gasoline also other little

Photo 235: Morris and our Dad, "Buff" Patteson.

Photo 236: Our Mom in a rare vanity pose - managed to get her 4 kids college degrees.

clean-up chores, fixing flats on trucks and cars. That was common in those days. You fixed flats nearly every day.

My mother was more or less self-educated but she did go to college and was an assistant instructor in English speech. She had enough credentials and experience to be a teacher. She came to Smiley as a school teacher and that's where my mother and father met and eventually married.

Getting back to the Naval air school, this new large base at Corpus Christie Naval Air Station not more than about 110 miles from my home town. I didn't have to make long travel arrangements to report for duty. So the Navy put you through the Navy primary flight school and they had plenty of room and plenty of airplanes. One incident that I recall was that I was with my instructor in one of the small biplanes in training and he knew that I had been in the Civilian Pilot Training Program and he says, "Do you know how to do a slow roll?" I said, "Oh yes, that was part of our training in CPT." And so he said, "Well, I don't know," the instructor, "Would you show me one?" So when we were up in the air I did a slow roll for him and I got a very good grade that day in my flight training.

The Navy had other things which I had already committed to memory and remembered about aviation. I knew quite a bit from the textbooks that I had in college, I knew about the weather, how to read weather maps and how to predict weather and the effects of weather

Photo 237: Early PBYs on patrol in Aleutians. Note radar antenna under wings.

and icing and so forth. I also knew a little bit about navigation and of course, the flight rules in aviation at that time were beginning to get complicated so I boned up on flight training. I had good grades for the Navy flight training. This went on until the final that I took in the Naval Flight School was, as I said, twin-engine float planes, the PBY. These I had not had any experience with before. They were large. They had a wing span of around 102 feet and a crew of six people, two pilots, a navigator, a plane captain, and two waist gunners. I always say the waist is the middle of the plane. So I went through that and got my wings in October of '42, 1942. Got three days' leave. Took my leave and went home and my orders had me back to duty and to report to San Diego. San Diego kept me one night and told me to report to Seattle. And to Seattle, I was told to report to Naval Air Station Whidbey Island, a brand new operational training base. And there they had the PBYs and I began operational training in the PBYs.

I had been at the Seaplane Base at Whidbey Island about 40 days and they said, "You're not going to be flying in PBYs anymore, we have a new aircraft that's going to replace the PBY. This is a land-based twin-engine aircraft made by Lockheed and called a "Ventura". I was assigned to a crew as a copilot navigator. The two junior officers alternated positions — one flight we would change from the copilot position, next we'd change to navigator's. And that went on until we got through training, which was a very short training because there was quite a bit to learn about this faster, hotter, twin-engine Lockheed fighter bomber. But they had a bomb bay to carry depth charges or bombs or an assortment of other pyrotechnics

Photo 238: Lockheed PV-1 Ventura #933 Training Flight near Victoria, Puget Sound.

that we had to learn about and this included torpedo drops. We trained in that. Our pilot's name was Birmingham. He had just come back from a tour in the Aleutian Islands in a P-boat. We were glad to have him because he was experienced in the weather and the living conditions that the Navy had to provide in the Aleutian Islands, which most of us had hardly ever heard of, the Aleutian Islands of Alaska.

Pat: I'm in Seattle and we were transferred up to Whidbey Island, which became a large Naval operational training base for not only patrol planes, and patrol bombers, but every other aircraft carrier. This was their final departure area to join a carrier and join the fleet. They left from Whidbey Island. They flew out to Puget Sound and made their carrier landing on the water and after they all got aboard, the dive bombers and the fighters and the scout planes, they just kept on sailing out to the Pacific. So we saw some of them, flew all around them but I never got on one. The Ventura and the PBY were too large. They were not designed for carrier landings. Too big.

After a very brief training for a very complicated airplane, the Ventura, the PV-1, we were set to go to the Aleutian Islands. Our first stop would be at Adak. En route we got to see some marvelous things. We flew together, two and three groups together until we got up to our first stop , a fuel stop and we were to stay overnight. This was one of the centers for a Native American tribe that inhabited those islands. The tribe was called the Tlingits. They were the indigenous group that had a large community. Our first night there we were invited to play basketball with them. Well, we didn't have a team but certainly we had enough young guys that were eager to get out and have some exercise. And Tlingit men and women were relatively short. I'm a tall man compared to what they were. Anyway, one night, the first night we were there, they sent a rusty old yellow school bus and picked us up, all that they could carry, some 20 of us, went to play against this high school native group. We looked them over and they looked us over. They had an auditorium there that was about the size of a large tennis court, not much larger, and one of the net goals was over a small stage. The other goal was nailed over the entrance door to the auditorium. There was just enough room for all of the squaws and other visitors, moms and dads, to encircle the court and stand with their heels next to the wall and their toes were actually touching the sidelines. We got a team together. Of course, most of the guys had some experience playing basketball in high school or college, but anyway, when they blew the whistle and the ball was thrown up for the jump-off, these fellows could jump as high as our fellows could reach and tipped the ball before it began to come down. And this kind of surprised us because all of their team didn't have uniforms, they had heavy shirts on and they had denim trousers, but mainly they wore rubber boots, every one of them, because the weather and the climate and the dampness and the rain and the snow, rubber boots was the thing everybody wore. Mom, dad, kids, and school children. All had rubber boots. Anyway, we thought we had an advantage over this basketball team because they certainly couldn't run as fast as we did. That was our mistake. They could run faster in those boots. It was loud, but it was faster and they ended up outscoring us about 2 to 1. So we had a good time and the squaws all were rooting for us, the Navy guys, which we thought was very accommodating.

Anyway, the next day was preparation to continue on up the chain and our next stop, I believe, was Anchorage or Kodiak. The next stop was up the chain, I forgot the station now, it was not well known. Finally we reached Adak. One by one we reached Adak and were immediately put into another school. A refresher school, identification school. That's where you learn how to identify a Japanese submarine from one of ours and learn how to identify their warships against our warships and identify their airplanes against our airplanes. So this was not the first but it was one of the most thorough training we had at Adak.

In the meantime, the Japanese had already occupied two of the Alaskan Islands, Kiska, and Attu. Attu was the end of the American territory and Kiska had been well fortified and had fighters and float planes on that base. So we were at Adak for maybe three or four weeks and flew patrols with our planes and this was out over the water. This was what they call a "sector search". You go out like it was a bicycle wire and you go out say 400 miles, you turn one way or the other, according to your orders, and you fly another 100 miles, which was across, and come back 400 miles. So this made about a 900-mile search. Shortly thereafter the fleet moved in and was going to retake the island of Attu. The Japanese had a garrison there of around three to four thousand troops and they had fortified it. They had their encampments and their equipment and a small airfield.

Our squadron was then sent to another island closer to Attu that was only about 120 miles from the Japanese at Attu and our new base. This was kind of roughing it, the first real rough living that we had. There was no running water. We were in tents at first. We had an oil stove in the tent and getting water or even to get a bath was almost impossible. You had to walk for miles to where there was a common shower room and most of us just stayed a little bit stinky and waited until they fixed a bath and shower room closer to our huts.

Photo 239: Amchitka Special - A monthly bath.

Photo 240: PV-1s head to the Aleutians.

The weather was not, at that time, very cold. It would frost over at night but it's not the severe cold that you'd think about the Arctic having. There was plenty of snow, but it melted usually about noon and made just puddles of water. And some of the guys went and drank that water. They gave us pills that would decontaminate the water. It tasted terrible but it was supposed to be healthy. I have a picture in which it shows me sitting in a wash tub and it was the first bath I'd had in a month. And that was done inside the warmth of the hut.

After that, we were transferred. In the meantime, the Battle of Attu had begun. This was a very difficult battle and these guys were in freezing weather a lot because they had to fight not just on the shoreline but the conflicts went all the way up into the mountains, higher mountains, which are three and four thousand feet up there and that's where the Japanese had fallen back and retreated back into the tops of the mountains. Now they suffered from it because they couldn't get food, very little. Our troops suffered because they were not really equipped for foul weather, mud, and ice and the rain continued all the time it seemed. So the battle for Attu lasted about twice as long as they had anticipated. The casualties were primarily frozen feet and frozen hands and just skin disease. So one of the main things that happened on there was that all these so-called people who had to turn into the hospital for various reasons of weather and chill and frozen fingers and so forth, they were attacked one night by a group of die-hard Japanese. It must had been something like, I don't remember, 70 maybe 200 Japanese, and they raided and killed all the people in the hospital who weren't armed of course. They finally were repulsed and our forces found them and there was a scrimmage and they were all killed.

So the Battle of Attu, there was not very much said about it. It was a fast one for battles as we later incurred in Europe, but it was one of the most severe and the percentage of casualties was one of the highest in all of the battles of the Pacific. This is when I was back from flying, you might say cover, for the landing force. Our squadron was the closest and we flew above the fleet landing forces to intercept any Japanese fighters or bombers that may try to intercept the landing forces. Actually, there were a couple of attempts but the weather was so bad that the Japanese bombers could not see the target to drop their bombs and we couldn't see or find them, except by radar, we could not find the interceptors. Of course, it was too far for their fighters to fly to Attu. The fighters couldn't make it but their bombers could make it.

This is when I and four other co-pilots, got orders to return to the United States. The squadron had to stay up there and they eventually took residence or occupied Attu since the Japanese were driven completely off. Four of us were put on DC-3s and sent back to Whidbey Island. They said that we had been up there long enough, three months or four months I think, and we were experienced, so they wanted to start another squadron with us. We had seen the land, we had engaged in the Japanese fighting and we had certainly endured the hardships and eaten all the SPAM that we wanted to and so we were good to be the core group of starting another Ventura PV-1 squadron. We were joined, however, by a group of ex-instructors. There are 18 crews to form a squadron in the Navy. We were joined by almost 14 new pilots. They were all senior to us. They all had instructor ratings but none of them had ever been in the PV-1 Ventura. So they had to learn to fly it. So we were acting instructors for them for a while, a short while, because they actually were more skilled at flying but they had been flying the old PBY, which was a different bird than the Ventura. At any rate, they caught on quick but we had very short training. We formed a new squadron and this time we had a better aircraft. It had better navigation. It had the LORAN. It had not only the navigation equipment but we had better crewmen. The gunners and the mechanics were better trained because they had more time to learn their trade.

Photo 241: U.S. Navy Pilot Pat Patteson

I became one of 18 crews. I became what they call a PPC. That's Patrol Plane Commander. I was 22 years old and I'm in charge of an airplane with a crew of six. We trained together at Whidbey Island, forming a new crew. This squadron was to replace the first squadron that had gone up and who had to learn everything from scratch. We were blessed by having some ex-instructors, very good at instrument flying which is essential in the Aleutians because the fog and other foul weather and storms made it necessary to be good at instrument flying.

That's where we flew from Amchitka and flew cover for the landing force that was going into Attu. It was dirty, muddy. Denise: *You flew from Amchitka?* Pat: You never heard of it? Neither had we. Amchitka put us about 140 miles the invasion and recapture of Attu. After Pearl Harbor, the Japanese sent part of their fleet up to bomb and attack Dutch Harbor in the Aleutians. Dutch Harbor was a military base but mainly it was for international fishing fleets. They were after the salmon during peacetime. The Japanese did not invade Dutch Harbor but they bombed it and bombed it heavily. They destroyed a lot of the buildings and some of the smaller ships and fishing boats that were in the harbor. At any rate, if you ever watch television and see these people that go out in the weather and get crabs and salmon and so forth, they headquarter at Dutch Harbor. At the time it was quite rustic and early and makeshift than it is now. It's quite a little commercial village now and maybe 40 years earlier, fishing fleets from Russia, Japan, Australia, the United States, and South America had headquarters there. They have a big canning operation during salmon season.

Editor's note: additional 36-minute interview and tour of Pat's house and WWII history files in his "War Room".

(All looking at beautiful watercolor paintings hanging on the wall)

Pat: My wife was an amateur artist. All the pictures in the house are some of hers.

She was pretty good. That's an old barn that was in Danville and that one, that's Ireland. She was a very creative person. My wife, from the effects of Alzheimer's, passed and gosh, that's terrible. Well, the reason we moved here at the Country Club is that she liked the view from the deck out there. And you can see Mt. Diablo looking this way. You can see the hill and the turkeys coming around promptly at dusk to go to their roost.

Denise: *So what was your best golf score ever?* Pat: I never got to single digits but I got down to say 14, 13, or 14 handicap. Denise: *Did you ever make a hole-in-one?* Pat: Yes, I'll show you. It's quite a celebration, yeah. I have a little plaque and the golf ball. It was in Houston. My brother-in-law and sister and their sons, we used to visit them in Houston and I forgot the little town out from Houston where they were but they were in a club and I always took my clubs with me when I went to see them. He was better than me but I don't think he had a hole in one. I got a hole-in-one on his course. He's not here with us but he was a very good golfer. I'm sure he got a hole-in-one but I don't remember. My wife took to golf in college but she was terrible (laughs). We never got beyond the second hole. Had to quit. She got some instruction at junior college near Pomona or someplace and she did very well, but they only taught her one club, the sand iron. It's the only club she could hit.

Here's my golf hole-in-one. Those are markers. I played at several courses and they give you this little marker to locate your ball on the green. You buy them. Some of them are, I went on a tour, or rather I put in for playing several courses in Europe and France and Scotland. I played the old course. I didn't do very well. If you're acquainted with the course or you have a good caddy, it's better. You do better. They more or less can tell you — not only give you the yardage and help and tell you what the greens are like and when the wind is coming up and all that, they can just give you encouragement. Denise: *Did you do bowling, too?* Pat: I used to bowl. Here in Danville.

Photo 242: As a Navy Cadet, I came home some weekends.

(Viewing aircraft models)

Lea: Well, this doctor used to have — his family is from Kamchatka. That was a place where Dad used to fly over, and he used to notice as a child these old planes were broken down. So he's a doctor. He lives in Florida now, but I gave him a little plastic model and he does such work, he takes five years to complete one of these models. So I put it behind glass to protect it. But that's actually a model of a plane dad flew and so he copied the decals and the number and the little painting on the side, so we're quite proud of that. Unfortunately, the little antennae looks broken now on the top but everything else is quite good. He made it. He brought it back. He flew up to visit us to spend the day for dad's birthday.

Pat: That's the PV-1. There are two models of the Lockheed, a PV-1 Ventura and a PV-2 Harpoon. The PBY is the flying boat over there. That's what I trained in. The Harpoon model is over here in the stockroom.

(Viewing collection of military records)

Pat: Up in that little bag is the original microfilm of every mission that my squadron went on and the briefing and the debriefing. The bomb loads and everything. I have probably the only record of everything that my squadron, which was VPB-135, that they ever flew. I have other things here, even a story of a Japanese Ace fighter that we encountered.

This says Myoshima, Shumushu, that's the Japanese air thing and this was on Shumushu. When this was a fresh picture, you could actually see three or four Japanese Zeros on that runway. But the film is faded out. Lea: *In Google Earth, you can still see the skeleton of a couple of the Zeros.*

Pat: These are aerial photographs that I took from my plane. This is a Zero. I have quite a bit of information on the Zeros. This is similar to the weather we saw. This is the big canning area on the Japanese island of Paramushiro. Hard to see is the roads and the little huts and tents and so forth. And there are some ships out here in the bay.

This is a Hiroyoshi Nishizawa. Nishizawa flew a Zero and we met [in combat] one time. On one of the photographic missions that I was on, the last air station of the Japanese was on Paramushiro and he and his group had been set up there urgently to defend the Japanese island and he and two of his wingmen intercepted us on our photographic run. I didn't know it at the time, of course. He has 86 kills. Denise: *So how did you beat this fellow here?* Pat: How did I beat him? Denise: *You came close to meeting him but then how did you for real meet him?*

Pat: I had a long photographic run and the last air defense that the Japanese had, we didn't fly directly over it, we flew alongside it and took pictures of it for intelligence purposes. And we were about 10,000 feet. I looked out the side window and there was a Japanese Zero there. There are two other Japanese Zeros on this side, on the left side, my side. And they had been reported by the crew that these guys were coming up to engage us and so I didn't know it at the time. I didn't know who they were, they were just Japanese fighter defenders. So my first impulse, and this has happened many times, we knew how the plane was fast enough that the Japanese Zeros could not repeat an attack. They could attack only once but then they had to climb back up, they lost altitude, the lost speed and they lost the capability of making an attack and so they had kind of taken our measure.

And here we are, both of us at 9 to 10,000 feet, two on one side and one on the other. Our navigator and gunners were trained on them, if they came in too close, within range, to open fire. They didn't have to because the Japanese stayed their distance. They kept out of range. So they followed us. By this time we had gone all the way past the target and were actually heading home. Our job was done. They started closing in and so I put my plane in a shallow dive and we got to speeds of what we knew was well over 300 miles an hour, which was pretty fast at that time. And the Japanese, that's pretty close to their limit. But anyway, the differential between our speed and theirs didn't allow them to make a typical effective firing run because to make a firing pass requires that you have overtaking speed of at least 50 to 100 miles an hour and their overtaking speed versus the plane we were in, in this dive, was probably less than 10 or 20 miles an hour. So we had an advantage. We could put more lead toward them than they could get into position to hit us. So what they do is, this is them, they get altitude and they make an S-turn and then come in and start deflection shot up here. Well, they couldn't do that. They come in here and they can't do this S-turn fast enough that we can't have as good an aim at them as they have at us and they kept their distance. They never really engaged us. I burned up all that altitude getting faster and faster and further away from the island and they gave up and came home. So that's how we met.

Nishizawa was killed in the Battle of the Philippines. They moved him and his squadron back to defend the Philippine Islands and some of the pilots that they were transferring, he was in a transport plane and the plane was shot down by some P-51s or P-38s, I don't know which. We found this out from the Japanese in Nishizawa's log book — they found his log book, on the day that we were up making our bombing run, our photographic run, he reported in his diary that day that he and his wingmen engaged with an American bomber, a bomber that had twin engines, twin tails, a turret on top, just like Mitchell's B-25. And he reported us as a B-25. We were a Lockheed plane. They are very similar when you count the bumps on the frog's back but nevertheless, he was mistaken in identifying us. They finally gave up and came back to their base and we eventually came on back into our base.

Pat: Well, we were talking about the microfilm in there but I have copies from the microfilm and I can show you some of that. Otherwise, I must say that the microfilm is in pretty bad shape. I won't show you the microfilm.

This other side has a lot of stuff about Alaska and I've got reports about the Kurils up there. These were three of our planes near ADAK in the Aleutians flying in formation. That's the Ventura. That's me in a Japanese foxhole. That's me with a plane. The plane was called "Pipsqueak." So was I sometimes. Can you find me in this picture? I'm the guy kneeling in front. That's me and this guy over here on the right was my copilot.

Denise: So are those all your medals and stuff over there?

Pat: Yeah. They all gave me that. Kind of a gift. A group, church group. And there's our patch. See that patch up in the middle top center? That's what you wear on your jacket. I've got pictures of me with that sewed on our jacket. In fact, I think that one right there has it. That's me and my copilot and we're in front of one of the planes, squadron planes.

(Viewing large collection of photographs, including many photos taken by Pat and another crew member who had a camera. Most of the photos are included in this book)

This is an early shot of the PBYs and that's the Malaspina Glacier. That's near Ketchikan, up in Glacier Bay, and all that area up there. That's what it looked like before the war. It doesn't look like that right now. The glacier has receded way back in. That's when it was extended out as far as it ever had been. This is a plane and actually, that was a plane that

I flew. This is a picture of planes in our squadron. The background there is snowcapped mountains I guess because they go up around three to four thousand feet. That was the airport, the airfield that we flew out of to go take our missions to the Japanese on the islands.

Pat: This is me over at Stockton in the Harpoon. The second version of that plane and the guy let me fly it two or three times.

Pat: Any of these photographs that you want, for instance, this is a photograph of the airfield on Attu when it was active and we were using it. These are two of the runways. I lived right there.

Pat: Well, the shadowbox up here with the medals. I have two DFCs, two air medals, the American Theater Ribbon, and the Asiatic Field ribbon with a star, which is the Attu Battle. And then there are two others associated with the Pacific activity that we were in.

Pat: These are patches that you sew onto your jacket breast for identification. This was my Lieutenant patches. This is a shot of the landing beach at Holtz Bay on Attu. It shows you the end of a little rivulet there. A couple of shots of it, one is of the early PV-1 and then this is the Catalina PBY-5A. 5A had wheels that could make it amphibious, land on land, or water. Here's an accident that happened on Amchitka. A P-38 came in and apparently, its nose gear failed. This is a seaplane tender. When P-boats landed in the harbor they didn't bring them up at this time, the early P-boats didn't have landing gear so they fueled them and got the provisions and changes for the crew by what they call a seaplane tender. This shows the act of trying to load a Mark 13 torpedo into one of the Venturas. Here is our skipper. He was Paul Stahl and he was, we thought, a great skipper. He was pretty stern but he was very easy to get along with. The other fellow is the Executive Officer, M.A. Mason. Here is a war correspondent and he's talking to Richard Rice. Richard Rice was my navigator and this fellow bummed a ride with us one time. He wanted to see what the enemy territory was like. This is the Lieutenant in charge of the Navy photo lab. They have a complete photographic laboratory there to develop pictures and we had aerial pictures.

This is one of the pilots, Frank Spiva. He lived in Abilene, Texas. This is me taking my monthly bath in a washtub. This is another pilot. This was a sunny day as you can see. One of the pilots, a pilot on one of the rare sunny days we had. This is a co-pilot. This is a pilot. This is my plane captain. He was a mechanic and I've loaned him my radio because the men, the crewmen didn't have a radio in their hut for entertainment. They had a small radio station set up at Attu eventually, months after we got there. You could tune it in and get a little bit of music.

Photo 243: 1944 Attu, Ens. Dick Rice with picture of his girlfriend Suzie, Minneapolis, MN.

Richard Rice, drinking beer. That was his milk of the day because he was from Milwaukee. Richard Rice with his girlfriend's picture. Her name was Suzie. This is my copilot in the plane. His name was Ray Walsh. More pictures inside the hut. This is my hut on Attu. Eight of us lived in that. These are just pilot shots. Two guys playing Acey-deucey. My copilot in his little bunk room inside the hut.

Up in Seattle, we went on a little snow

Photo 244: 1944 - Me and Lt.(jg) Holloway play acey-deucy (backgammon) on our "off day" schedule.

trip one time and these fellows are playing with a couple of husky pups. Picture of the plane. Picture of the snow in Attu. Picture of the assistant navigator we used from time to time. And this is aboard the Coast Guard frigate. One time going back to the States we had to go at the end of our first deployment. I didn't fly back, I came back on a smaller plane. This is a shot of part of the landscape of Attu. This is Adak.

This is in my college days when I was taking civilian pilot training and this is the fellow that was teaching me in Abilene, Texas. This is me when I was a cadet with a cadet uniform on and my twin brother, Morris. Here are my two sisters and a cousin when they were in school. Here's my brother, twin brother. This is my twin brother when he was in med school in Galveston, Texas, I gave him money one time because he wanted an engagement ring to get engaged. The girl turned him down so he bought a sailboat and I got to ride in the sailboat twice. (Laughs). It was really my sailboat.

This is a postcard of Seattle before the war. This is a picture of the beautiful BOQ when we first got to Seattle, we almost had a private room. You shared it with somebody else. This is the Deception Pass Bridge, built over the small inlet between Whidbey Island and Fidalgo Island. On this side of the bridge is the inland waters of the Puget Sound.

More family photos, I've got to get a photo of my mother out here. This is my mother, me, my brother and my father. He was a Naval medical student. These are my two sisters who sent me a Christmas card. This is Nelda and this is Beth. This is my wife, Jolea, The Artist. And this is on a fishing trip someplace. This is aunts and uncles with my dad. This is the four children of my family. This is some old tiny pictures of a great-grandmother or someone I never heard of.

This is on Attu. We had this ramshackle old school bus that was sent up to us and we used it for taking us from our quarters out to the planes for a mission. This is seven of us. We had an extra man there who was an assistant navigator because they thought we needed — that's before LORAN. We had to use celestial navigation and you needed somebody that knew a little bit about celestial navigation.

These are some of the early decorations. These are some of my hut mates. This was a specialist for Lockheed. He was paid by Lockheed but he came up and was a specialist on the mechanical workings of the Ventura. We needed that. This was a copilot. This was my copilot. That's me. Christmas time and another pilot went after a Christmas tree.

Some beautiful cloud shots made by the photo lab. I don't know why I have them. It didn't have a beautiful day.

This gets into some crews. This is Stepter's crew and we're in front of PV-2s. This is another crew. Holloway was a native of Forth Worth I think. Spiva's crew and he, after the war, came back and became a geologist and lived in Abilene, Texas. This is Stepter's crew. This is Westus' crew. I don't know the crew members' names. This is Pearl's crew. This fellow is from Corpus Christie, Holland. This is Brownie's crew and there are just some extra men here. These were two chiefs, Chief of Aviation Machinists.

This is Vivian's crew. He was forced to go into Russia and spent several months there. That's a story in itself. There have been books written about these internees that had to go into Russia. It was only about 150 miles from our target area. From the Japanese Islands to the Russian Kamchatka Peninsula.

Rumford's crew. He was a funny guy. Always laughing and joking. This was part of the assistants and he was a gunnery officer. This is shots outside, me, shots outside.

Our Executive Officer and you see he's got the patch on there. Here are a couple of guys looking at pictures beside the photo lab. They had a huge photo lab there. Employed about a dozen guys. Here was a sunshiny day and a bunch of the guys just sitting around shooting the bull. Here are some fellows working on an engine outside. This is Rumford, my copilot, and Nordbie, I'll tell you more about him later. He put out a little pamphlet for the squadron telling jokes and what was going on and when the next dance is going to be. A newsletter, it was. I've got a whole book of them. We had a fellow in the squadron that was a cartoonist. Pretty good too. Okay. This is a bunch of guys. This is one of our planes on the revetment. The people are on a little knoll there watching and waiting for planes to return.

This is Ketchikan, which was primarily had been a fishing port. This is the Port of Kiska Island, which was occupied by Japan. This is after they were forced off and that's a ship that's run aground there. Here is it.

This is a kitchen. This is where they cooked up all that delicious food. I got acquainted with the man that runs the Alaska Aviation Heritage Museum in Anchorage and I gave him some stuff. I found some relics over on the battlefield of Attu and turned them over to him. One was a mess kit that still had rice in the bottom of it (laughs). And another was our astronautical clock. Chronometer they called them in those days.

This is a friend. He came back after the war and got a job with the FAA, Federal Aviation Administration over here and I went to see him one time. He was in the FAA office in Oakland.

Here are some things that are in that museum up in Anchorage. This is one of the crewmen that came back and he got a job in this FAA station.

Now, you can write a book about the first lady on the land. The Admiral thought his crew, his assistants needed fresh milk. They were tired of powdered milk and it tasted like powder too. So he thought that it would be a good idea since he had one of those CP plane tenders that needed to have an overhaul and had to go to Bremerton to get it done, that's in Washington, "Why don't you just bring back a fresh cow and we will have fresh milk?" You didn't grow up on a dairy did you? So he took a collection from his staff, he had a staff of about eight guys, he took up a collection enough to buy a cow and bring her back on this boat that went down to be overhauled.

Pat: They built her a barn. Here's the tender over here. They gave the cow a barn. She had a telephone and running hot water in that barn. They actually brought back some chickens to make the cow feel at home. The problem was, they would graze her on the weeds. The weeds were bitter. Her milk became bitter and nobody would drink it. The Admiral, turned her over so the fresh milk could be put in the officer's bar so they had fresh milk instead of a highball or mixed drink. And nobody would drink it because it was bitter. So they finally sold her to an Army unit that butchered her and had hamburgers from then on.

This was probably before your time. This was one of the few entertainers that came up to the Aleutians. Most of them went to sunnier places. This is Joe E. Brown. Ever heard of that name? He was an entertainer. He was the star in some old movies. I won't go back to silent days but I mean it was almost that far back. Anyway, he came, they brought him back up there and he's looking out the window of a PBY. Here's one of our planes being fueled up.

Pat: Picture of a plane. I think that's me in there. In the background, oh here it is, you used to be able to see a city back there. That's Victoria, British Colombia. This is Puget Sound underneath there. We were training at Whidbey Island.

Whidbey Island. PBY. Some of Alaska's beautiful scenery. Here's a snow plow. Here's one of our planes. Here's an area of Holtz Bay in Attu. PBY. Here's the Admiral giving me a decoration. He gave me one of those ribbons in there.

This is a control tower on Attu. This is me at the cockpit. This plane crashed somewhere. That was not uncommon for planes to crash up there. My third crew. My plane captain, my radioman and my gunner, turret gunner and my copilot down here. Now this is Holtz Bay and this is one of the seaplane tenders that stayed there and these are some of the little scout planes that flew around there.

My turret gunner sitting in the turret. My plane captain. Picture of the navigator and the navigator's table and the turret gunner and the mechanic on a mission. This is taken inside the plane while it was flying. I took the picture with the little camera I had. This is the lieutenant that was in charge of the photo lab. Here's another, same picture I think. PBY. This is Attu. You can see one of the runways here.

Pat: I took most of the photos. This is, in the third deployment, this was my navigator. This is me in the plane on oxygen, taking oxygen. We were flying very high to get over a storm or something. This is Amchitka, an aerial shot. My second crew, my third crew. They used to give out these little cards, I've got mine in there. The dark spots are night missions and the red bombs are day missions. This is a colorized picture of the plane. My copilot and me. Myself. Plane mechanic. Assistant navigator. Plane mechanic. Copilot. I took all these.

This is Rice, the second squadron, Rice, the guy from Milwaukee. This is just a cutup. He decided he would paint three stars on his helmet so he would look like a general. This is my third crew. This is my copilot, me, and my navigator. These are early postcards of Kodiak, Alaska, before the war. I bought a bunch of them. Here's a husky. This is Main Street of Kodiak, Alaska. This is a Russian church. This is a scan of a brown bear. This is a bank in Kodiak. This kid's having lunch. Back in the old whaling days, this was a whale that the natives killed. This is an early polar bear hunt, brown bear hunt. This is a cache where you keep your food up away from the bears. Canoe. These are all early postcard stuff.

This is my navigator. I think you saw this before. That's a war correspondent. This is my navigator and his wife who came up one summer to see us. They lived in Ohio and he went back and became a farmer after the war.

Rumford, who the cutup was. Here's a guy fishing for salmon. Here they are again. Here's some salmon they caught. Here's my crew, their decorations, the decorations of my crew. They look different in their Navy dress blues. Radioman, turret gunner, mechanic, assistant, and navigator. This is on Puget Sound and that's a ferry. It goes between the island and Seattle island. This is the guy who was my roommate for a while early in the war and he and his pilot crashed over here in Alameda and were killed. This is a pilot and his copilot and this is his navigator and my navigator. They just mixed them up. They were having a day out in the sun. I took most of these pictures here. I didn't take that one.

This is Attu. This is where we flew from and went 650 miles to the target. Runways are here. I lived there. The hanger area was right over here. It's hard to see. These pictures are kind of faded out. But this is Holtz Bay. It was an inland bay. The fellow that was killed, that was his burial flowers. This is on Adak. PBY plane. This is Cookie. Cookie adorned the bar. She was the background. Actually, it was drawn by memory because we never had anybody like that (laughs). It was drawn by memory by a fellow that was not in the squadron. He was a Seabee. Construction Battalion. He was good. This is another ship that ran aground over the inner bay. That's the same picture, same ship. And a cow barn, and our crew. And this is a scout plane. It lands on water.

Photo 245: Cookie

My navigator, my assistant navigator, and my mechanic. The Coast Guard ship that I came back on from one time, and this is coming through the inland passes. I took all these. That was outside of Ketchikan. I think that's me. This is one of the ship's navigators. He's taking a shot of the sun. This is the skipper of that ship. Kind of a rough individual. It's about as good as I could say. He got along fine with us. We were his passengers, see, so he tried to make us feel at home. This is at Ketchikan. You can't tell the tide at Ketchikan was at high tide was here, down here was 12 feet down before you got to water again.

Photo 246: Snowqualmie Pass Ski Resort, WA.

Pat: You know, in the bay out here we get a tide two feet. This is a Snoqualmie Pass. This lady took three of us while we were in Seattle up and tried to teach us to ski. She did a good job but we didn't. Me, a couple of guys, we were on Lake Washington in Puget Sound and we were having a day off and we rented a boat. A cousin. A fellow that was in my hometown. This is the girlfriend of the guy that got killed. This is his sister and their dog. This dog is a Great Dane. They lived in St. Louis so I didn't get to take the pictures. This is a family named Benge. This girl was in high school. This boy was going into high school. This is her graduating from high school. This is her mother. This is another baby they had. The oldest one was a man who went through Navy training like we did but he opted to get into the Marines and become a Marine pilot. Me, my brother. My brother was a med student. Here he is in his med student uniform. My brother was in a men's club at school in Galveston. Their names are here. They had a party and this is me at home. This is me and a girlfriend, another pilot, and his girlfriend in Seattle. This is another friend in Seattle. This is a celebration of Navy dentists. One of them had a son, his first son, that's his wife and they're celebrating by getting drunk. The son's name was Donald. I didn't know any dentists, but I thought that picture was good.

Here's some of us at a nightclub in Hollywood, me and my girlfriend, Tom and his girlfriend. Mitzy. She was a ballet dancer. Here's a bunch of us in our dress blues at the BOQ in Whidbey Island. We got together. We were all single. So on Whidbey Island somebody had to arrange a party or there wasn't anything. This is a shot of the Whidbey Island landing strip from my window. Here I'm writing home. This is a bear. This is another bear somewhere. There he is. He's barely visible. On the Komandorski Island. This is a Russian Island right off the Kamchatka Peninsula. We're hitchhiking a ride. Here we went over to Canada. This was a car four of us bought. It's an Oldsmobile and we all bought it. Each of us pitched in 100 bucks and bought that car. Here's a picture of the gang. The bunch of single guys. My copilot and so forth. These are just newspaper clippings and so forth. I had a lot of them. We sent newspapers back home. They don't even make good reading. Here's the Admiral pinning a medal on me and here's three of us that got my Smiley Flyer Much Decorated for Feats Against Japanese. How about that, huh? This is a map. Maps are not very thrilling but it shows where we were. We were out here. This is the end. This, then going back, then going back, then going back all the way. It's 1000 miles from here to here. The mainland of Alaska is here. The Alaska Peninsula starts and the islands over out there. Some of them are just rocks sticking out of the ground. Another map. This is on Attu but that's kind of dull. That's where we lived.

Denise: If I remember right, I could be wrong, you got taken away, sent back to Seattle and you had to teach these seasoned pilots how to fly the Ventura.

Pat: Yeah, the old fellows, they were three or four years older than we were sometimes. We had to show them a few things but they were good at other things. They were, for one thing, they out-ranked us. We were just barely Lieutenant JGs and they were all full Lieutenants. But they were nice guys and most of them were married and they found lodging for their wives up at Whidbey Island. That was the second deployment when these ex-instructors came into our squadron.

Pat: ...Things are kind of calmed down between my birthday and getting kicked off the carrier. They were going to have a big show in Hawaii and they invited, the invitation went out for all the WWII veterans to be there and they would treat them royally and so forth and they were going to take a lot of the WWII vintage aircraft on the carrier over to Hawaii and we were supposed to load these planes down at San Diego. Well, to get on the carrier, the Navy has been in trouble. One of their ships caught on fire down in San Diego through carelessness or something. Anyway, the Navy was rather tender about a lot of things. So when we got on the carrier they gave us all a physical exam to see if we were carrying COVID-19 and sure enough, well I wasn't carrying it but my blood pressure was high. So they kicked me off and everybody else, not everybody else went but I didn't get to go to Hawaii and have fun. Well, I was disappointed at first but we were going to fly this old bomber from Stockton down to San Diego and we get just about to Los Angeles and they had a huge oil leak on one of the engines and had to make an emergency landing. So the Navy sent a car to pick us up and take us down to San Diego. So the first night I got to spend on the carrier and I got a very swanky stateroom.

Denise: So did you take most of those pictures yourself in all your books or was it a combination?

Pat: A lot of them I did. There was strict censorship on this and cameras were not supposed to be, private cameras were not supposed to be used up there. The Navy could use them. I sneaked an Argus C3 35mm. And there's another guy in here. He and I teamed up and he had a camera, about the same kind, and we took a little thing that you could load the film with and chemicals and so forth. We hid them in our airbags. I took pictures but censorship was supposed to eliminate that. You were not supposed to do that. As you can see there's not very much to hide up there. We got to see some things that no one else had the opportunity to see at that time. One of the areas of Alaska, shown on the aerial maps, had a place called the Valley of Ten Thousand Smokes. It started with a volcanic action in 1912 that overflowed and spouted all this dust and pulp or whatever comes out of a volcano and it covered a small stream. This pumice was hot and it was about, over this stream, it was about 15 feet thick. So the pumice stayed hot for years and caused steam to rise up, hundreds of them, little spouts of steam. I flew over that. It doesn't exist now. It's not active anymore. It spouted steam for 40 or 50 years. 40 years at least.

Denise: So I see you've got a nice quilt over there. Did they present that to you, the quilt?

Pat: Yeah. I went to a birthday party. Somebody invited a bunch of guests, friends of mine and theirs, including Vietnam Veteran Gil Ferrey, and she just arranged that for this quilt with someone in the group. There's a description of it over there. But there's a group that makes them. I guess they sell them. And you can buy them from several ladies that provide this quilt dedicated to veterans. I got a lot to learn from them. So I do a little speaking with others. It was started by pilots in WWI and they drank together and ate together after the war was over and they came home and they thought they ought to stay together.

Denise: Well, it's kind of a life-altering experience to share together. That is the first war they had aircraft.

Pat continues the interview...

Pat: We had three deployments to the Aleutians. The second was, of course, the longest one, and this one involved missions to the northern islands of Japan which were the Kurils. We were to get photographs of the installations and any other intelligence information we visually had. We were equipped with the PV-1 and it had cameras that were aerial cameras for taking shots at altitudes around 10,000 feet.

I was at one time designated to be the photo plane that had five different cameras, three in the nose and two in the tail. There was a diversionary group of about six planes from Attu that were to coordinate and attacked the fighter base on Shumushu, which is the northern island of the Kurils. They carried this off very well and was partly a surprise to the Japanese, however, we lost two planes. They had to be taken to the Kamchatka Peninsula because they were shot up and had some wounded aboard. My plane coordinated with them and I was to turn five cameras on for vertical and oblique shots of the airfield at Shumushu, then down to Takamatsu, and then down to two other fields in Paramushiro. We more or less ran the gauntlet of fighters because by this time the fighters had taken off from their main field on Shumushu and were following us looking for an opportunity to coordinate a fighter attack on us. However, our speed with the Ventura was so comparable to the Japanese fighters that we were able to stand them off successfully and we were hardly hit by any of their guns. We finally ended up the flight at the southern part of Paramushiro and were able to escape from any bad damage and return to Attu, our base.

It was revealed later on in peacetime that one of the fighters at the third airfield had three aircraft come up to intercept us. Two of them came up to our port side and would make faking runs at us and that drew our fire to the left side of the plane and the third Zero would then come in and make a fighter attack with the intent of damaging us. This fighter plane happened to be piloted by one of the most outstanding fighters that the Japanese had, a fellow by the name of Nishizawa. He was a Lt. JG in the Japanese Naval Forces and he was to be the one that was going to be credited for a kill. However, because we dove and had such speed that the closing rate was sufficient to keep them from completing their attack. So we got away with hardly any damage and got home safely with something like over 300 very valuable aerial photographs of five airfields on the Kuril Islands.

Post-War, Marriage, Retirement

Editor's note: Follow-up interview with Pat & Lea.

Pat: I think a mention of the business I got into after the war. It was about the time of this man on the moon with John Glenn and I had a small business in Livermore that was helped by people that understood high vacuum pumps and other related equipment because the high vacuum, of course, is to simulate the conditions of outer space. This business I had was to repair pumps used to create a chamber that was devoid of any atmospheric gases such as you find on the moon or in outer space. I had technicians that knew how to repair this equipment and I set up a business in Livermore. The business finally became a small business and I asked my son who had a job in Los Angeles if he could change jobs and come and help me and he readily accepted because I think he was not overly enjoyed by his present work. So he came and joined me and became my outside sales representative and our business grew and we were able to take on more work with companies that were involved with space equipment, particularly the astronauts and various equipment used in space travel.

So that kept on and we were in the business for maybe 14 or 15 years in Livermore until unfortunately, my son had a stroke. It incapacitated his whole right side, particularly the use of his legs and right arm. So since he was the primary worker beside me in the business, we had to quit the business and retire. I was of retire-able age and he was very close to it. Besides, his presence was very essential to keeping the business together and running smoothly. That took care of our business, which I think we had something like 15 or 17 years of business.

Pat: Yeah, the business started in '81 and when did it stop?

Lea: Five years ago [2017].

Pat: Five to seven years ago, yeah. So since then we've retired comfortably and live together in a house in Danville and we make some visitations to the air shows. We enjoy that together. The Veterans organizations have helped us meet a lot of new people in their clubs and organizations, which has included some trips to other air shows and my capability of flying the PV-2 owned by Taigh Ramey in Stockton.

The other activities we have together have been more or less local and still we are entertained by some of the organizations, Veteran organizations, and this has been most helpful to keeping us delightfully busy with friends, local and far away.

Denise: Okay. Is that all you wanted to say?

Pat: I want to mention my wife. We met after the war. After World War I had a job with a freight airline in San Antonio, Texas and she also was working in San Antonio as a secretary. We met through a mutual friend, also a worker at the airport service in San Antonio, and we learned that we both came from the same county. She was from Gonzalez, the county seat, I was from Smiley, a small farm and ranch village in the same county. We married in Gonzalez, Texas. I had a job with this freight line and I moved around quite a bit but together we ended up in Burbank. I became a flight controller to dispatch and route the freight airlines for pick-up freight. Finally, we fell in love with California to the extent that I ended up being moved to Northern California, particularly Palo Alto where there was a freight station. Eventually, I left Palo Alto, my freight airline job in Palo Alto, I left it to start my own business, which was the repair business that I mentioned.

Denise: So after WWII you left the military?

Pat: I had left the military when the war ceased. I was still on Attu and I had enough, you might say Victory points, to be separated from the military and I put in for separation, which was granted. I was given the chore to leave Attu and fly one of the old bomber planes all the way back to Abilene, Texas and then take a bus home by gratis of the Navy. I continued to stay with some Naval Reserve training courses but being moved around so much by the freight airline, I had positions that included Dallas and St. Louis and Chicago and Dayton and Cleveland that I finally decided that I would pick Palo Alto as my final move and that's when I got into my own business.

Denise: Okay and what was your rank when you left the Navy?

Pat: I had become a Lieutenant Senior Grade, Full Lieutenant sometimes referred to. So that's the equivalent of the Army's Captain, two silver bars. That was the last military command that I had.

Denise: How did you keep your pilot's license up? Did you have your own plane?

Pat: Actually, the Navy and the FAA allowed us to be, without charge, given an FAA rating. I ended with a rating that was single and multi-engine land and seaplanes so I could fly anything except a balloon. I kept it up. To sustain your rating, your flight rating and so forth you do have to have so many hours in the type of plane. I found this was difficult to do, in fact almost impossible after I got into my own business, to stay into any type of a flying job. So I lost my ratings but I did keep up with them in reserve training for almost two years and then finally had to stop. I couldn't make the meetings. I couldn't keep up the flight time. Besides, now I was a civilian and it cost money to keep up with those things. I did have my own plane for a while. I enjoyed flying a little four-seater Mooney. I had a Mooney M20F and it was a single engine. I enjoyed very many trips and my family went with me to Mexico and to Dallas and San Antonio and back to our native homes and flew to reunions in Reno. Squadron reunions were held nearly every other year. One of them was at the training field at Ault Field in Washington, the State of Washington. So that was an enjoyable time. I was not the only owner, it was a partnership ownership is what I had. But I enjoyed the little plane. It was very fast and very convenient for that matter. We had a partnership. I and another couple. We had to sign up several weeks ahead of time. We were responsible for all the maintenance and fuel that the plane needed on our trips and we shared a certain amount of common maintenance for the examinations and ratings that were required by the FAA. I did it for about four years. Just an estimate. But we went everywhere we wanted to. Denver was another stop. Denise: *Did your wife enjoy flying with you?* Pat: I tell you, I think she did. She didn't know very much about it. We went back to the manufacturing site which had kind of a ceremony and homecoming for their pilots and owners and they had an organization that was going to teach the wives how to do an emergency landing. This was very fine except that my wife flunked (laughs). Denise: *I hope there was no emergency.* Pat: For that reason, we didn't have any emergencies (laughs). Denise: *No emergencies allowed.* Pat: Right. I think otherwise she enjoyed the expedition. In other words, if we were to go back to her hometown of Gonzalez, Texas, it would take us nearly three days. If we flew in the little airplane, we could do it in eight hours. She liked that part of it. And my son was involved in it later on but he was not licensed then when he went to school at Cal Poly in San Luis Obispo, I insisted that he take flying lessons and he got his private license eventually. He couldn't keep up with the requirements of practice and so his license is not up to date.

Denise: So you said that after the war you started working for which company again?

Pat: The company where I got my first job, I came back from the war and I had at least two more years of college to get a degree in engineering. When that happened, the company I was with was just beginning to turn into an air freight organization. Their name was Slick Airways, Inc. The name Slick was the owner's name. There was a family that had a background in oil drilling so they became active in that until they finally ran into a problem with the government. The government would not grant them a postal permit to carry mail. This is a very lucrative license that major airlines used and they could carry mail and freight for the postal service and receive quite a good compensation for that. Slick Airways and another airline could not obtain a license to carry mail. The other airfreight company was Flying Tigers Organization. They all had old C-46s that had been used in the war and most of these pilots that they used were ex-military pilots who flew the C-46 on the hump in China. The India-Burma military equipment and personnel were carried across India into China by going over the Himalayas, which they called "The Hump." I didn't pilot with them. I wanted a ground job. I was married so I was looking for something a little safer and with a better, I thought a better future. And it turned out neither one of those things happened. I finally got out and got into my own business, which I described.

Denise: So what did you do for Slick Air Freight?

Pat: I started out as a freight handler. I wanted to get my foot in the door and kind of learn from the ground up and after I graduated from the University of Texas they hired me. The only position they had open was you could be a freight handler. This means you actually load and unload an airplane and put it in the trucks for delivery. Later on, in just a matter of a few weeks, they began to expand a little bit and I became an Assistant Station Manager while the regular station managers at Dallas and St. Louis, and Chicago took their vacations. They hadn't had any vacation so I had a job for about three weeks in those different places.

Later on, they expanded some more and I was offered a Station Manager position in Dayton, Ohio. The station manager that was then in Chicago, excuse me, Cleveland, the station manager in Cleveland, which was a larger freight station, he decided he didn't want that job anymore, he wanted to go back to his former job and they offered me that position and the last job in Ohio was as Cleveland Station Manager. From there I was then promoted to a flight controller position in Burbank and that's how I came to be in California. I came to Palo Alto after the flight control job because they had an opening for another station manager here. So I liked the station manager position in as much as it seemed more, well, my wife and I both liked it better.

Denise: So what happened to Slick Air Freight? Did they just finally go under? Did it get bought out?

Pat: They went under and they did both of those. They went out of business, went under, and they sold the airplanes. Eventually, the airplanes became obsolete because instead of two-engine planes now they went into four-engine planes such as the Boeing Douglas DC-4. The planes became antiquated and they couldn't afford to buy the new four-engine cargo planes.

Denise: Okay so what was your favorite plane to fly? Which one did you enjoy the most?

Pat: Well in the military I liked the Ventura best. In civilian life, I was able to buy in on this little Mooney. That was a fun plane. You could make a lot of little short trips, land closer to the center of the town and it was a simple but very fast little plane. It was almost as fast as the bomber. So it was quite good. But it was a simple plane but very easy, simple and very easy to fly. Much easier than military planes.

Denise: So you keep in contact with the people you knew back in Alaska?

Pat: Actually not anymore. We had reunions when the war broke out, about five or six years after peacetime, after the war. We had our first reunion was going to be in Las Vegas and we had two or three more reunions, one in Los Angeles, one in Seattle, one in Denver, and Las Vegas. We had four or five reunions. I still have a roster of the personnel but unfortunately, they have all passed. Most that I knew have passed. So I don't have any close correspondence with any of them now.

Denise: So did you ever get back to Alaska as a civilian?

Pat: I have. I've been up there on a fishing trip and a sightseeing trip. My wife and I went up on a sightseeing trip and I've gone by myself on a fishing trip back to the area of Kodiak. Actually, it was on the mainland but the big place we landed was Fairbanks and then Kodiak. So I have seen some of the same territory I've seen before. But not in the Aleutians. I've not been to the Aleutians again.

Denise: So I guess Alaska hadn't changed much then, right?

Pat: No, it hadn't changed. 50,000 people don't live on those islands anymore. Some of them are totally uninhabited now and just have government visits like Coast Guard or various protection agencies, the National Park Service keeps them up. Some of the villages, the native villages that existed before have been reinhabited and grown and others have become commercialized such as the salmon and crab fishing areas out at Dutch Harbor. You can see some of the [difficult] conditions, commercial conditions are still shown on television with the small crab boats going out and fishing.

Other than that, I understand that Attu, where we spent most of our operational time, is almost totally uninhabited. The natives have not recovered their village on Attu. In fact, of the about 28 natives that were incarcerated by the Japanese when they took it over, most of them have passed on now. I noticed a report on that. The Kiska area is almost uninhabited. Coast Guard goes out there and checks on radio and navigational aids and other little jobs of Coast Guard duties for lights and safety and other navigational aids that shipping uses. They still have to be marked. They still had to be checked for electrical and space applications. They get involved with relay work of telephones and other things. One harbor that is still very active, as I mentioned, is Dutch Harbor. It still remains a very big commercial fishing, crabbing, and fishing port.

Denise: Where did you go in Alaska to go fishing?

Pat: The fishing trip was around Kodiak and let's see, Anchorage. That was in the Anchorage area. I went trout fishing and I went up there by myself but I stayed in a camping area so we had shacks and so forth, all the conveniences as a sportsman type thing. That was fun. Stayed a week up there. Caught some trout. It was very pretty. Also saw a lot of bears and eagles and everything like that so

Photo 247: Fishing.

it was a fun trip. There are plenty of animals still there. I think the number of black bears has gone up. The number of Kodiak bears has stayed about the same. Probably the biggest bargain in the traveling business for vacationers, not necessarily fishing, is the Inside Passage. If you ever make a tour or travel log and want that kind of experience, you see more beautiful things and enjoy more animals, and have an easier ride, is to go the inside passage from Seattle or Vancouver all the way up to Anchorage or Denali, the big mountain up there. My son's been on it. He's been up there too. But that, to me, is probably the biggest bargain and most exciting of trips. It's also reasonable, relatively. You don't have to go to Europe to see anything. You get on a boat here in Oakland or Seattle or Vancouver or someplace like that and be there. It's an easy trip and it's smooth. You don't have the big swells of the ocean. Denise: *So you really like Alaska?* Pat: Yes.

Denise: *So did you ever go to Japan?* Pat: I've never been to Japan. The closest I ever got was about 10,000 feet. I would like to go. I would like to go and I have a lot of Japanese friends now. Most of them are US citizens and they are respectful and probably born here and so forth. The hatred and disgust we had for the way the war started and the people and the military, particularly the military personnel were ruthless many times and I never appreciated that. Fortunately, none of the Ventura squadron planes were ever shot down over enemy territory. That was fortunate because they didn't like aviators and they usually had some pretty drastic means of showing their disgust for us. I won't tell you what they did but it wasn't very nice. ...I've been to Europe two or three times. I've traveled to France and Spain and Italy and Germany, and England.

Don: Before I let you go, last question, what's your favorite plane? If right now you won the lottery, what plane would you go get? You only could get one though.

Pat: Well, if I won the lottery, which is up to a billion dollars now, isn't it? One ticket gets you a billion and something dollars. I don't really know. I would probably opt for a business jet. They're beautiful. Leer has set the high mark for building a good plane others are probably better now and of course, they're all equipped with satellite equipment. You push a button and they tell you exactly where you are. It doesn't take a major airport for them to land. You can build, if you had that much money, you could build your own airport.

Pat Patteson Associates, Inc.

Editors note: This 45-minute interview was recorded 19NOV2021 at Pat's 100th Birthday Party. I had just met Pat and we were sitting with a few other guests before the party started.

Pat: ...at least two years of college to qualify at 20 years old. So I went back to school. I went back to the University of Texas and after that, I did a hell of a job trying to find a job in flying. All those positions were taken as there were so many pilots. And the airlines wanted four engine stuff and I only had two engine stuff so they got the first crack at it and first options. I wrote to about four or five airlines and never got a reply from any of them. So I figured that's a dead dog. (Laughter) But I finally got into my own business. I had been transferred. I was in with an air freight company and I got transferred to Burbank and from Burbank, they put me into the flight control for that little freight line and I ended up in Palo Alto. I've been here for 20 years now.

Lea: You got out of the airline industry after a while, pretty soon after that. We were a stocking distributor of laboratory pumps so that's what we've been doing.

Pat: Yeah, our business was -- and Lea and I had a business of repair on the test chambers and so forth for the space program which was just breaking open and so forth. I helped with the engineering and proposals for Lockheed's environmental test chambers and so forth like that. It was interesting and at that time it was the thing, you know. Space was the thing. So I got in kind of on the ground floor so far as the equipment and testing of it is concerned.

Don: Hi Pat. I'm Don Downey. I'm friends with Don Long, the Pearl Harbor survivor.

Pat: I know him very well.

Don: You really got my attention Pat, when you said you went to the University of Texas. If you would do me a big favor and pick up the story there.

Pat: I finished with an engineering degree and looked around for a job and some of these surplus airplanes like the C-46 and so forth were available and several people tried to start their own airlines, particularly in freight. That was an area that seemed to be appropriate and they picked up these C47s, that Curtis twin-engine freight carrier. I worked as a station manager for St. Louis and then I went to Chicago and then I went to Cleveland and then I went to Dayton and so forth. Finally, I ended up in Burbank as part of the flight control unit for this airline. Flying Tigers and Slick Airways were the two major freight carriers that were trying to get started. Unfortunately, those had freight contracts for hauling the mail, all of those contracts were taken and that was the lifeblood of airlines at that time.

Finally, I was lucky to get a job with a company that makes scientific equipment, Bell and Howell. I got with Bell and Howell and from Bell and Howell, I knew a lot about the maintenance and servicing and clean-up of vacuum pumps. The vacuum pumps led me into the environmental chamber business and I worked in that and, as I mentioned, Lockheed, I was in Palo Alto at the time and Lockheed was building these huge chambers to test the equipment for the first astronaut flight. Everything from the wristwatch to the capsules had to go through an environmental test program for temperature and heat and be able to withstand the take-off of a rocket. They had vibration tests and so forth. So they also wanted to get the test equipment of real life. In other words, you don't just test a small item but you test the whole capsule and so forth at the time, for its job in space. From there I picked up some people that were, or I knew people that were familiar with vacuum and vacuum equipment, high vacuum such as the atmosphere of the moon, and so my son and I started a business with that and that's where I ended up finally quitting the business and retiring.

The business, vacuum equipment for working with these test chambers, there was no one or very few people, that knew how to repair them and clean them up after they had been in a test or a failure test and so I found that some of the big companies that were in bidding the contract, they had to send the vacuum equipment back to the manufacturer. In some cases, this was back to Germany. So it was easy to gather some of the individuals who were kind of tired of doing everything themselves and it was a small company, maybe eight or 10 of us, got collected. So I started my own business and put my own name on it so there you are. After that all you had to worry about is government regulations. No problem (laughs). That continued, I guess from, what did we have it for, about 25 years. And I had a shop location out in Livermore. I retired five years ago I believe, my son will correct me on all this.

Lea: Well, we started in '81. Then I had a stroke about five years ago which knocked me out of the business and dad, being that he was 95, figured that he might want to retire.

Pat: Yeah I stayed with the business for quite a while. And as Leasays, his inability to -- he was the sales department, the trucking department, and some office work but when he became incapacitated due to the stroke, at my age I said, "Why keep fighting it? I'm going to get out of this thing." And I was successful enough to get out of it and retire comfortably.

M: Let me ask a question here. If my math is right, you started that business when you were about 70. And you were working on what is really the edge of technology. So the experience and the knowledge to be on the frontiers of science there in the space program… It was a vital part. People who were going to explore space got to be ready to work in a vacuum.

Pat: Yeah, I was old enough to be an astronaut but (laughs). That's what intrigued me. John Glenn and I probably are very close to the same age. But I never attempted an application with them. I had my business and so forth and it was going along. I never really got serious about whether I wanted to try to be an astronaut or not. Well, it was obviously a new adventure for everybody and I was, of course, fascinated but then my part in it was more or less picking up the pieces and putting them back together again (laughs). I attended to preparing big proposals for the company and wrote up the technical proposal for a couple of large chambers. Not by myself, but I mean I participated in it.

Houston was just starting up the flight, space center there. But some of the equipment that we designed and our expertise was in high vacuum work. In other words, we had to know what kind of pumps could evacuate air and all the other stray gases to simulate space. And the Lockheed guys, I remember the first conference table we sat around with, they didn't realize that air was just -- their engineers learned from us what once you get beyond 40 or 50,000 feet things change and they become items that required specialized pumps. There were electronic pumps. There were mostly at that time there were diffusion pumps. And of course, the mechanical pumps were simply getting the space chamber up to maybe 75,000 feet, but now you've got space that changes. You're not -- the molecules of air that we have right now you can actually feel them. They're a fluid. And later on they simply become molecules like bees in a room. They're all going at different ways at different speeds and you have to be able to either ionize them or in some cases a diffusion pump, which is a combination of ultrasonic speeds of oil and the Lockheed engineers who were in charge of getting this equipment, this huge chamber -- it was eight feet in diameter and about 40 feet high and it's still there, the old Lockheed. It was getting this equipment engineered for what we knew that the conditions of the moon. We knew that. So that's what you had to, that was an example of what you had to attain. That performance is what you had to attain.

Don: I bet the astronauts were thankful to think when they closed that hatch, right?

Pat: Well I've been to a space museum in Washington but that was several years ago and they had the original capsule there that Shepard and John Glenn and, I think Shepard was the first guy in space. He just went up and came down. John Glenn had three trips around and so forth.

Pat: My association with Lockheed goes right to their planes. I flew the Lockheed planes in World War II. And got familiar with some of their engineers but our job was to -- I wanted to stay in multiengine work and I got into these Venturas and Harpoons which was twin-engine stuff. I was in the first Lockheed Ventura Squadron that went to the Aleutians. So it was quite a transition because before then all my training had been in PBYs. Later on they made them amphibious but for the first ones, you had to pull them up a ramp by hand to get them on shore. All of that took place in Corpus Christie. Many of us didn't like the idea of where we were assigned because the Aleutian Islands are not a fun place and I was up there for three tours. We'd come back, reform, and go back up. Three times, '43, '44, and '45.

Pat: It was beautiful. It was so barren. It's all volcanic. The islands there were created out of volcanic eruptions and it's part of the Circle of Fire. And of course, they had a woman under every tree (laughs) but there weren't any trees. Somebody took a 2 x 4 or 4 x 4, drilled some holes in it, built up some weeds and stuck it in there, and made a Christmas tree out of it. That was one of the first trees I ever saw there (laughs).

Pat: Aeronautically we brought in some of the first planes, land-based bombers, and patrol planes. We went from the PBY which was pretty good at cruising about 105 knots and it also landed pretty close to that. But that was one of the most interesting planes that you can imagine. It had a crew of around seven or eight. They had the first radar. We got it from Brittan for patrol work. And later on we got the first Loran for navigation. Before then they gave us an octant or a sextant to do our navigation with instruments of that type but when we got to the Aleutians you never could see the stars anyway (laughs), or most of the time you couldn't see the stars. But it was an education. The plane was so adaptable that one of the items in the plane was a little tin horn and when you landed in a harbor or something or where the fog or something was a problem and you didn't want to run into any ships in there, someone had to go, one of the crew had to go and get in the nose of the PBY and blow that horn every few seconds. Just like you were a ship coming in, you know? The PBY probably fascinated a lot of us because it was so adaptable. Obviously, it had quite a bit of action six months after Pearl Harbor, Dutch Harbor was bombed and the PBY always came out second best in any kind of a contact. They just couldn't keep up with the Japanese fighters.

Don: They painted them black and then they'd go in at night sometimes.

Pat: The Black Cat. There was a Black Cat Squadron but ours had the traditional shades of blue for camouflage. So we didn't have any Black Cats although the instrument flying was uppermost in the conditions of Alaska and the Aleutian Islands and the Kamchatka Peninsula where we ended up sometimes.

Don: Tell me about Texas. How did you wind up in Texas? Were you raised in Texas?

Pat: Oh yes. My hometown was a little farm and ranch community named Smiley about 60 miles southeast of San Antonio. I was farm and ranch. It's rolling and it had a good climate, I mean, rainfall and so forth. It was nice for grazing and the cattle. It was also fertile enough so that you could grow crops on it, corn and hay and so forth could be grown on it. I came through the depression years it was still going on. And somebody asked me how was that? Well, I was quite young. It lasted until the war started. But nobody starved. I mean we took care of our neighbors and so forth. The cattle and the hogs were butchered for meat. There was a meat market and restaurant and so forth. But nobody had any money. There was a bank there but nobody had anything in their pocket to rattle, you know?

Don: For me, I always thought it was hardest on the dads. The kids, maybe you guys got to get by but I just can't imagine being a father and not being able to get a job or just that whole thing I thought was not the best of times.

Pat: That happened. Not in my family necessarily. I mean we were average in terms of food and clothing and shelter. But a lot of them weren't.

Don: And then how did you get in the military? How did you make that transition?

Pat: Well to get in you had to be 20 years old. You had to pass the physical. You had to have at least half the credits toward a degree and they didn't tell you what kind of degree you had to have. You could be quilting if you want. You could have a degree in quilting and still get into the Navy or the Air Force. So at the time of Pearl Harbor I had already signed up on my birthday, which is in November, and Pearl Harbor happened December 7, so I was signed up and had taken my physical but I didn't have any orders to report anywhere. And I didn't get any orders to report until January 2. That was about three weeks after Pearl Harbor. You go through Christmas wondering when the telegram's going to come, you know?

M: Being from Texas, what made you want to go into the Navy?

Pat: Accidental. Partly. Well, my father was an Army Lieutenant in World War I so I was kind of expected to stay in the Army. However, I was in school, in college and a friend of mine, a classmate, came through. I got to talking to him and he had on that beautiful green Marine flight uniform and had the gold wings up here and I had never seen a green uniform before in my life. But you might say that kind of gave me the edge to go to the Navy. So I went down and within a few days of turning 20 I went down to a recruiting office which was in the basement of the post office. And I got to talking to this Yeoman and he gave me all of the application information and so I filled it out and sure enough they called me up and gave me a bus ticket to Dallas to get my physical examination. So from there on, that sign they put across your drawers when they issue your underwear, "Property of the US Navy." You stayed that way (laughs).

M: They stamped it on your forehead.

Pat: It was stamped across your ass.

Gil: Are you Pat? I'm Gil Ferrey, how are you? I was not in the Navy. I was in the Army flying helicopters in Vietnam but a friend of yours, Taigh Ramey's assistant Melissa Morehead called an organization, the Commemorative Air Force, and I got a call and said, "Can you put together some kind of a gathering for Pat on his 100th birthday?" and I thought well, I think we can probably bring three other 100-year olds or a couple 99-year olds. [discussion of the recent passing of Pearl Harbor Survivor Don Long]

Pat: I knew about Don Long too, by the way. They brought a PBY to Napa, some of the people returning from the 75th anniversary in Hawaii and they came by and stopped to pick up Don Long at Napa. I was invited to go with them and we went over and took off in this PBY off the runway at Napa in the Catalina. And we flew over to Lake Berryessa and they landed on the water, floated around a while, and took off again and came back to Napa and put the wheels down instead of the floats. I was very interested. I told the pilots that that's what I had trained in.

Gil: You flew up in Alaska quite a bit? Out of Adak or where? Or Attu?

Pat: At first it was Adak and then it was Amchitka and then it was Attu. Attu mostly started in 1944 after the Kiska Garrison had been evacuated. They evacuated the Kiska Garrison and Attu, the battle of Attu I was involved. I was a copilot then but I was involved in it.

Gil: Were you there when there were P-38s in the 11th Air Force up there and they flew with a B-17 mothership to navigate because the fog was so bad. And one of them, six of these P-38s took off one day from Attu, I think from Attu, and they flew toward Paramushiro in the Kuril Islands and they intercepted some Japanese betty bombers coming in to bomb Attu. And when the Betty Bombers saw these P-38s who were a little more maneuverable than the Betty Bombers, they turned tail and headed back to Japan. But what we didn't know was that they had since mounted machine guns in the tail of the Betty Bombers so as they turned tail, they fired at the P-38s and shot down the squadron leader, never seen again, and another pilot, who I met, that had to land his P-38 in Massacre Bay (Attu). And they picked him up with a Kingfisher. An observer, I guess that's what he was called. He had a pilot and an observer in a Kingfisher. He got out of his seat, walked out on the wing and popped the Mae West of the surviving pilot of the P-38, pulled him onto the wing, put him in his seat and they taxied back to a tender, poured hot coffee down both ends of him, gave him a hot shower and he survived.

Pat: I never knew any details like that but I do remember the incident of them intercepting the Betty Bombers. I've assembled a lot of information. I have all of the sortie reports, the briefing reports from the pilots and crew for every sortie that our squadron was in. We had four Ventura Squadrons up there and I was in one squadron. I had three tours.

Gil: Was the Ventura the name of the first model that Lockheed built and there was a second model called something else?

Pat: That was the original, yes. The Harpoon came out in 1944. It was more of a patrol plane. It has more stability. Less speed. The Ventura, we liked it. Some pilots liked it and some pilots were not really up to speed with it you know? They had been trained over Alaska and the Aleutian Islands in PBYs and they just did not want to accept this frisky little plane instead of that docile float plane that had so many creature comforts.

Gil: What speed could you cruise at in the Ventura?

Pat: Well, for long-range cruises or missions, we generally cruised at one might say an economical cruise and it was about 150 knots. That's about 170 miles an hour.

Gil: Was it set up to drop torpedoes?

Pat: Heaven help me I don't want to drop a torpedo. In a practice. I dropped one accidentally. The people back in New Hampshire at the torpedo center. They just didn't make anything that could withstand -- the PBY could drop it okay. They're flying at 100 knots. The Ventura, we wanted to be at 200 knots when we dropped. The torpedoes couldn't handle it. They flipped. They went down and just made a splash and you had to get right down the throat of the guns to be accurate with it anyway.

Gil: Did you drop gravity bombs also?

Pat: Oh, yeah. It's a long story but I was in three squadrons and we all did different things each time. We had photo flash bombs at first because they didn't have any good photographs of the Kuril Islands and the best things they had in reconnoitering was, what's this magazine, National Geographic. They used National Geographic to learn what was on those islands. And that got outdated quickly.

Gil: I love reading books written by all the men. I've got a bunch of the 101st Airborne, 82nd Airborne, D-Day, and Iwo Jima. I got a chance to meet Dizzy. He was only 16 when he received the Medal of Honor. He lied about his age and forged his signature. Little guy. He was only about 5'5". He jumped on a hand grenade at Iwo Jima. And then about a week later he jumps on a second-hand grenade, this one's not a dud, to save the guys in his foxhole. It blows a lot of his stomach out. So he had about four operations. He spoke at a lot of different events. A little short fiery guy. I majored in history but I've got a big World War II library. I've just always enjoyed World War II history. I do other types of military history but World War II is my favorite area. That's why I enjoyed going over and walking the battlefields a lot.

(Introductions to birthday party WWII Veterans)

Gil: This is Diz Laird (WWII Ace Navy Pilot Diz Laird) and Lou Gibbs over here who flew Navy B-24s or PB4Y-1s. The two came later. Single tail, the Privateer. That came later. You flew the double tail. That was stolen from the Army Air Corps of course. And this guy (WWII Veteran Jake Larson) landed on D-Day in Normandy.

(End of audio)

References

Abandoned & Little-Known Airfields: Alaska, http://www.airfields-freeman.com/index.htm, http://www.airfields-freeman.com/AK/Airfields_AK.htm#amchitka, © 2003, © 2022 by Paul Freeman. Revised 1/14/22 [26AUG2022]

Air Classics "Venturas of the Empire Express - Part One" by Boris Ilchenko February 2017 (Magazine Article)

Air Classics "Venturas of the Empire Express - Part Two" by Boris Ilchenko March 2017 (Magazine article)

Aleutian Islands World War II National Historic Area, Unalaska/Dutch Harbor, AK, https://www.nps.gov/aleu/index.htm [28AUG2022]

Florida Warbirds, https://floridawarbirds.com/2021/05/06/naval-aircraft-on-display-at-orlando-sanford-international-airport/ [29JAN2023]

Pacific Wrecks, Paramushiro Island. https://pacificwrecks.com/provinces/russia_paramushir.html, Pacific Wrecks Incorporated is a non-profit public charity 501(c)(3), [28AUG2022] and https://pacificwrecks.com/aircraft/pv-1/33346.html (Photo by Ralph Wetterhorn 2002)

Perry-Castañeda Library (PCL) Map Collection World War II Maps website, https://maps.lib.utexas.edu/maps/historical/history_ww2.html [26AUG2022]

Video of Pat Patteson's interviews can be found on YouTube, https://www.youtube.com/channel/UCUto-iOpDLNYoJmmaCFz8fg

World War II Database, https://ww2db.com, C. Peter Chen, The United States Library of Congress recognizes WW2DB as a valuable online resource for its researchers. WW2DB is filed under the ID lcwaN0005634, [26AUG2022]

Glossary

Airport apron, Taxiway, Runway. (27AUG2022) https://en.wikipedia.org/wiki/Tarmac

Barnstormers: Barnstorming was a form of entertainment in which stunt pilots performed tricks individually or in groups that were called flying circuses. Devised to "impress people with the skill of pilots and the sturdiness of planes,"it became popular in the United States during the Roaring Twenties. Barnstormers were pilots who flew throughout the country to sell airplane rides and perform stunts. Charles Lindbergh first began flying as a barnstormer.Barnstorming was the first major form of civil aviation in the history of aviation. (27AUG2022) https://en.wikipedia.org/wiki/Barnstorming

CAVU: Ceiling and Visibility Unlimited, an aviation meteorology term. (27AUG2022) https://en.wikipedia.org/wiki/CAVU

Celestial navigation: also known as astronavigation, is the practice of position fixing using stars and other celestial bodies that enables a navigator to accurately determine their actual current physical position in space (or on the surface of the Earth) without having to rely solely on estimated positional calculations, commonly known as "dead reckoning", made in the absence of satellite navigation or other similar modern electronic or digital positioning means. Celestial navigation uses "sights", or timed angular measurements, taken typically between a celestial body (e.g. the Sun, the Moon, a planet, or a star) and the visible horizon. Celestial navigation can also take advantage of measurements between celestial bodies without reference to the Earth horizon, such as when the Moon and other selected bodies are used in the practice called "lunars" or lunar distance method, used for determining precise time when time is unknown. Celestial navigation by taking sights of the Sun and the horizon whilst on the surface of the Earth is commonly used, providing various methods of determining position, one of which is the popular and simple method called "noon sight navigation"—being a single observation of the exact altitude of the Sun and the exact time of that altitude (known as "local noon")—the highest point of the Sun above the horizon from the position of the observer in any single day. This angular observation combined with knowing its simultaneous precise time referred to the time at the prime meridian directly renders a latitude and longitude fix at the time and place of the observation by simple mathematical reduction. The Moon, a planet, Polaris, or one of the 57 other navigational stars whose coordinates are tabulated in any of the published nautical almanac or air almanacs can also accomplish this same goal. (27AUG2022) https://en.wikipedia.org/wiki/Celestial_navigation

chandelle: The chandelle is an aircraft control maneuver where the pilot combines a 180° turn with a climb. It is now required for attaining a commercial flight certificate in many countries. The Federal Aviation Administration in the United States requires such training. The chandelle (which is the French word for candle) is a precision aircraft control maneuver, and not strictly speaking an aerobatic, dogfighting, or aerial combat maneuver, however it was used with success by Japanese Zero pilots of the Tainan Air Group in 1942 over New Guinea. It is rather a maneuver designed to show the pilot's proficiency in controlling the aircraft while performing a minimum radius climbing turn at a constant rate of turn (expressed usually in degrees per second) through a 180° change of heading, arriving at the new reciprocal heading at an airspeed in the "slow-flight" regime, very near the aerodynamic stall. The aircraft can be flown in "slow-flight" after establishing the new heading, or normal cruise flight may be resumed, depending upon the purposes of the exercise or examination. See the diagram for a visual depiction of how the maneuver must be flown for the purpose of certification. The pilot enters a chandelle at a pre-determined airspeed in the normal cruising range for the aircraft. To begin the maneuver the pilot first rolls the aircraft in the desired direction with the controls (the ailerons), and quickly but smoothly establishes a medium-banked turn. In most small aircraft (cruising speeds of 100–175 KIAS) this bank will be about 30° to 40°. This will begin a turn of the aircraft in the direction of bank. Simultaneously, full power is applied and a smooth pitch up is started with the controls (the elevators on the empennage). The angle of bank stays constant during the first 90° of the change of heading, while the pitch angle increases steadily. At the 90° point in the change of heading, the aircraft has the maximum pitch angle (which should be close to the critical angle of attack at the level stall speed of the aircraft). During the second 90° of the change of heading, the pitch angle is held constant, while the bank angle is smoothly decreased to reach 0° of bank, the end of the turn and return to straight-and-level flight at exactly the reciprocal heading (180° away from the heading at the start of the maneuver), and with the airspeed close to the stall speed. The aircraft should not lose altitude during the last part of the maneuver, nor during the recovery, when engine power may be used to re-establish normal cruising speed on the new heading. The decreasing bank angle together with the decreasing airspeed during the second half of the chandelle will maintain a constant turn rate. The turn needs to be kept coordinated by applying the correct amount of rudder throughout the maneuver. From a practical point of view, the chandelle may be used to turn an aircraft within a minimal turn radius. As such it is a useful maneuver for pilots of small aircraft who find themselves in a blind valley or canyon. It was also, therefore, a useful maneuver to early fighter pilots in their low-powered aircraft to quickly turn toward a pursuing attacker (which would tend to make a tracking gunshot more difficult because of the turn and climb involved) while climbing but not stalling the aircraft, or to position themselves quickly to make an attack on a turning enemy or an enemy flying on another heading. French aviators during World War I described it as *monter en chandelle*, or "to climb vertically". However, it has since come to have a rather more strict definition for pilot flight testing purposes. In early fighter combat, a pilot would aim to achieve the maximum practical speed at full throttle, then roll the aircraft to the required angle of bank for the change of heading that was required in the combat situation. As the turn began, the nose is increasingly pitched up, as required for the amount or change of heading desired. As described above, the turn progresses with the desired pitch angle and requisite rudder to control yaw, airspeed decreases, and then the wings are leveled so as to roll out straight and level on the new desired heading, all the while avoiding an aerodynamic stall. At this point, if the pilot has any more power, it may be added to regain airspeed, or the nose of the aircraft can be allowed to drop somewhat to achieve the same effect (dropping the nose of the aircraft is not permitted for the purposes of the maneuver in commercial flight testing). In aerial combat, the chandelle maneuver was used both aggressively to position the aircraft for attack, and defensively to evade an enemy.

Commemorative Air Force: The CAF was founded to acquire, restore and preserve in flying condition a complete collection of combat aircraft which were flown by all military services of the United States, and selected aircraft of other nations, for the education and enjoyment of present and future generations of Americans. More than just a collection of airworthy warplanes from the past, the CAF's fleet of historic aircraft, known as the CAF Ghost Squadron, recreate, remind and reinforce the lessons learned from the defining moments in American military aviation history, (29AUG2022). https://commemorativeairforce.org/

diffusion pumps: Diffusion pumps use a high speed jet of vapor to direct gas molecules in the pump throat down into the bottom of the pump and out the exhaust. They were the first type of high vacuum pumps operating in the regime of free molecular flow, where the movement of the gas molecules can be better understood as diffusion than by conventional fluid dynamics. Invented in 1915 by Wolfgang Gaede, he named it a diffusion pump since his design was based on the finding that gas cannot diffuse against the vapor stream, but will be carried with it to the exhaust.However, the principle of operation might be more precisely described as gas-jet pump, since diffusion plays a role also in other high vacuum pumps. In modern textbooks, the diffusion pump is categorized as a momentum transfer pump. The diffusion pump is widely used in both industrial and research applications. Most modern diffusion pumps use silicone oil or polyphenyl ethers as the working fluid. (29AUG2022) https://en.wikipedia.org/wiki/Diffusion_pump

Dog' day: The dog days or dog days of summer are the hot, sultry days of summer. They were historically the period following the heliacal rising of the star system Sirius (known colloquially as the "Dog Star"), which Hellenistic astrology connected with heat, drought, sudden thunderstorms, lethargy, fever, mad dogs, and bad luck. They are now taken to be the hottest, most uncomfortable part of summer in the Northern Hemisphere. (27AUG2022) https://en.wikipedia.org/wiki/Dog_days

ETE: estimated time en route

Flying Tigers: The First American Volunteer Group (AVG) of the Republic of China Air Force, nicknamed the Flying Tigers, was formed to help oppose the Japanese invasion of China. Operating in 1941–1942, it was composed of pilots from the United States Army Air Corps (USAAC), Navy (USN), and Marine Corps (USMC), and was commanded by Claire Lee Chennault. Their Curtis P-40B Warhawk aircraft, marked with Chinese colors, flew under American control. Recruited under President Franklin Roosevelt's authority before Pearl Harbor, their mission was to bomb Japan and defend the Republic of China, but many delays meant the AVG first flew in combat after the US and Japan declared war. The group consisted of three fighter squadrons of around 30 aircraft each that trained in Burma before the American entry into World War II to defend the Republic of China against Japanese forces. The AVG were officially members of the Republic of China Air Force. The group had contracts with salaries ranging from $250 a month for a mechanic to $750 for a squadron commander, roughly three times what they had been making in the U.S. forces. While it accepted some civilian volunteers for its headquarters and ground crew, the AVG recruited most of its staff from the U.S. military. The Flying Tigers began to arrive in China in April 1941. The group first saw combat on 20 December 1941, 12 days after Pearl Harbor (local time). It demonstrated innovative tactical victories when the news in the U.S. was filled with little more than stories of defeat at the hands of the Japanese forces, and achieved such notable success during the lowest period of the war for both the U.S. and the Allied Forces, as to give hope to America that it might eventually defeat Japan. AVG pilots earned official credit and received combat bonuses for destroying 296 enemy aircraft, while losing only 14 pilots in combat.The combat records of the AVG still exist and researchers have found them credible.On 4 July 1942 the AVG was disbanded and replaced by the 23rd Fighter Group of the United States Army Air Forces, which was later absorbed into the U.S. Fourteenth Air Force with General Chennault as commander. The 23rd FG went on to achieve similar combat success, while retaining the nose art on the left-over P-40s. (27AUG2022) https://en.wikipedia.org/wiki/Flying_Tigers

hole in one: In golf, a hole in one or hole-in-one (also known as an ace, mostly in American English) occurs when a ball hit from a tee to start a hole finishes in the cup. A ball hit from a tee following a lost ball, out-of-bounds, or water hazard is not a hole-in-one due to the application of a stroke penalty. (27AUG2022) https://en.wikipedia.org/wiki/Hole_in_one

IFF: Identification, friend or foe (IFF) is an identification system designed for command and control. It uses a transponder that listens for an interrogation signal and then sends a response that identifies the broadcaster. IFF systems usually use radar frequencies, but other electromagnetic frequencies, radio or infrared, may be used.It enables military and civilian air traffic control interrogation systems to identify aircraft, vehicles or forces as friendly and to determine their bearing and range from the interrogator. IFF is used by both military and civilian aircraft. IFF was first developed during World War II, with the arrival of radar, and several friendly fire incidents. IFF can only positively identify friendly aircraft or other forces.[2][3][4]If an IFF interrogation receives no reply or an invalid reply, the object is not positively identified as foe; friendly forces may not properly reply to IFF for various reasons such as equipment malfunction, and parties in the area not involved in the combat, such as civilian airliners, will not be equipped with IFF. IFF is a tool within the broader military action of Combat Identification (CID), the characterization of objects detected in the field of combat sufficiently accurately to support operational decisions. The broadest characterization is that of friend, enemy, neutral, or unknown. CID not only can reduce friendly fire incidents, but also contributes to overall tactical decision-making. (27AUG2022) https://en.wikipedia.org/wiki/Identification_friend_or_foe

Inside Passage: The Inside Passage (French: Passage Intérieur) is a coastal route for ships and boats along a network of passages which weave through the islands on the Pacific Northwest coast of the North American Fjordland. The route extends from south-eastern Alaska in the United States, through western British Columbia in Canada, to northwestern Washington state in the United States. Ships using the route can avoid some of the bad weather in the open ocean and may visit some of the many isolated communities along the route. The Inside Passage is heavily travelled by cruise ships, freighters, tugs with tows, fishing craft, pleasure craft, and ships of the Alaska Marine Highway, BC Ferries, and Washington State Ferries systems. Coast Guard vessels of both Canada and the United States patrol and transit in the Passage. The term "Inside Passage" is also often used to refer to the ocean and islands around the passage itself. (29AUG2022) https://en.wikipedia.org/wiki/Inside_Passage

Jane Plane: navigation plane leading group of planes.

Judas goat: A Judas goat is a trained goat used in general animal herding. The Judas goat is trained to associate with sheep or cattle, leading them to a specific destination. In stockyards, a Judas goat will lead sheep to slaughter, while its own life is spared. Judas goats are also used to lead other animals to specific pens and onto trucks. They have fallen out of use in recent times, but can still be found in various smaller slaughterhouses in some parts of the world, as well as conservation projects. Cattle herders may use a Judas steer to serve the same purpose as a Judas goat. The technique, and the term, originated from cattle drives in the United States in the 1800s. The term is a reference to Judas Iscariot, an apostle of Jesus Christ who betrayed Jesus in the Bible. (27AUG2022) https://en.wikipedia.org/wiki/Judas_goat

LORAN: LORAN, short for long range navigation,[a] was a hyperbolic radio navigation system developed in the United States during World War II. It was similar to the UK's Gee system but operated at lower frequencies in order to provide an improved range up to 1,500 miles (2,400 km) with an accuracy of tens of miles. It was first used for ship convoys crossing the Atlantic Ocean, and then by long-range patrol aircraft, but found its main use on the ships and aircraft operating in the Pacific theater during World War II. LORAN, in its original form, was an expensive system to implement, requiring a cathode ray tube (CRT) display. This limited use to the military and large commercial users. Automated receivers became available in the 1950s, but the same improved electronics also opened the possibility of new systems with higher accuracy. The U.S. Navy began development of Loran-B, which offered accuracy on the order of a few tens of feet, but ran into significant technical problems. The U.S. Air Force worked on a different concept, Cyclan, which the Navy took over as Loran-C, which offered longer range than LORAN and accuracy of hundreds of feet. The U.S. Coast Guard took over operations of both systems in 1958. In spite of the dramatically improved performance of Loran-C, LORAN, now known as Loran-A (or "Standard LORAN"), would become much more popular during this period. This was due largely to the large numbers of surplus Loran-A units released from the Navy as ships and aircraft replaced their sets with Loran-C. The widespread introduction of inexpensive microelectronics during the 1960s caused Loran-C receivers to drop in price dramatically, and Loran-A use began to rapidly decline. Loran-A was dismantled starting in the 1970s; it remained active in North America until 1980 and the rest of the world until 1985. A Japanese chain remained on the air until 9 May 1997, and a Chinese chain was still listed as active as of 2000. Loran-A used two frequency bands, at 1.85 and 1.95 MHz. These same frequencies were used by radio amateurs, in the amateur radio 160-meter band,and amateur operators were under strict rules to operate at reduced power levels to avoid interference; depending on their location and distance to the shore, U.S. operators were limited to maximums of 200 to 500 watts during the day and 50 to 200 watts at night.(27AUG2022) https://en.wikipedia.org/wiki/LORAN

Mark 13 torpedo: The Mark 13 torpedo was the U.S. Navy's most common aerial torpedo of World War II. It was the first American torpedo to be originally designed for launching from aircraft only.They were also used on PT boats. (27AUG2022) https://en.wikipedia.org/wiki/Mark_13_torpedo

microfiche: A microfiche is a sheet of flat film, 105 × 148 mm in size, the same size as the international standard for paper size ISO A6. It carries a matrix of micro images. All microfiche are read with their text parallel to the long side of the fiche. Frames may be landscape or portrait in orientation. Along the top of the fiche a title may be recorded for visual identification. Microforms are scaled-down reproductions of documents, typically either films or paper, made for the purposes of transmission, storage, reading, and printing. Microform images are commonly reduced to about 4% or one twenty-fifth of the original document size. For special purposes, greater optical reductions may be used. Three formats are common: microfilm (reels), microfiche (flat sheets), and aperture cards. Microcards, also known as "micro-opaques", a format no longer produced, were similar to microfiche, but printed on cardboard rather than photographic film. (27AUG2022) https://en.wikipedia.org/wiki/Microform#Microfiche

Norden bombsight: The Norden Mk. XV, known as the Norden M series in U.S. Army service, is a bombsight that was used by the United States Army Air Forces (USAAF) and the United States Navy during World War II, and the United States Air Force in the Korean and the Vietnam Wars. It was an early tachometric design that directly measured the aircraft's ground speed and direction, which older bombsights could only estimate with lengthy manual procedures. The Norden further improved on older designs by using an analog computer that continuously recalculated the bomb's impact point based on changing flight conditions, and an autopilot that reacted quickly and accurately to changes in the wind or other effects. Together, these features promised unprecedented accuracy for daytime bombing from high altitudes. During prewar testing the Norden demonstrated a circular error probable (CEP)[a] of 75 feet (23 m),[b] an astonishing performance for that period. This precision would enable direct attacks on ships, factories, and other point targets. Both the Navy and the USAAF saw it as a means to conduct successful high-altitude bombing. For example, an invasion fleet could be destroyed long before it could reach U.S. shores. To protect these advantages, the Norden was granted the utmost secrecy well into the war, and was part of a production effort on a similar scale to the Manhattan Project: the overall cost (both R&D and production) was $1.1 billion, as much as 2/3 of the latter or over a quarter of the production cost of all B-17 bombers.The Norden was not as secret as believed; both the British SABS and German Lotfernrohr 7 worked on similar principles, and details of the Norden had been passed to Germany even before the war started. Under combat conditions the Norden did not achieve its expected precision, yielding an average CEP in 1943 of 1,200 feet (370 m) (a CEP of 1200 feet means 16% of all bombs dropped land within 1000 feet of the target), similar to other Allied and German results. Both the Navy and Air Forces had to give up using pinpoint attacks. The Navy turned to dive bombing and skip bombing to attack ships, while the Air Forces developed the lead bomber procedure to improve accuracy, and adopted area bombing techniques for ever-larger groups of aircraft. Nevertheless, the Norden's reputation as a pin-point device endured, due in no small part to Norden's own advertising of the device after secrecy was reduced late in the war. The Norden saw reduced use in the post–World War II period after radar-based targeting was introduced, but the need for accurate daytime attacks kept it in service, especially during the Korean War. The last combat use of the Norden was in the U.S. Navy's VO-67 squadron, which used it to drop sensors onto the Ho Chi Minh Trail in 1967. The Norden remains one of the best-known bombsights. (27AUG2022) https://en.wikipedia.org/wiki/Norden_bombsight

seaplane tender: A seaplane tender is a boat or ship that supports the operation of seaplanes. Some of these vessels, known as seaplane carriers, could not only carry seaplanes but also provided all the facilities needed for their operation; these ships are regarded by some as the first aircraft carriers and appeared just before the First World War. (27AUG2022) https://en.wikipedia.org/wiki/Seaplane_tender

Shit on a Shingle: In the United States, chipped beef on toast was commonly served to service members of the United States Armed Forces from World War I through Vietnam. It was considered emblematic of the military experience and consequently also served to Boy Scouts of the era.In American military slang it is commonly referred to by the dysphemism "Shit On a Shingle" (SOS). (27AUG2022) https://en.wikipedia.org/wiki/Chipped_beef#U.S._military_cuisine

sitting duck: A sitting duck is an easy target; someone who is easy to attack or criticize; someone in a very vulnerable position; someone easily caught or found. (27AUG2022) https://www.idioms.online/sitting-duck/

Slick Airways: Slick Airways was a cargo airline from the United States, that operated scheduled and chartered flights between 1946 and 1966. The airline was founded by Earl Slick, a Texas aviator and multimillionaire who along with his brother had inherited $25 million (around $324 million in 2015 currency) after their father's death in 1930. The airline was formed in January 1946 as the air cargo division of the Slick Corporation, headquartered in San Antonio. Slick Airways had its original fleet of Curtiss C-46 Commando aircraft based at Lockheed Air Terminal (Burbank) and San Francisco Airport. In 1949, scheduled freighter flights to domestic destinations were commenced, and by 1951, the company had become the largest all-cargo airline of the United States. On 16 April of that year, Slick Airways became the first airline to operate the freighter variant of the Douglas DC-6 (the passenger variant had been introduced with United Airlines five days earlier). In 1954, Slick Airways went into merger talks with Flying Tigers because of the increasing competition by passenger airlines, which failed because the respective shareholders would not approve of these plans. In 1958, the company was forced to suspend all scheduled services, though it continued to operate charter flights on behalf of the United States Armed Forces, also to European destinations. A fleet of newly built Lockheed L-1049 Super Constellation airliners went into service with the airline during 1959, initially being used to fly military freight from Travis Air Force Base near Sacramento to Japan. This operation was supplemented by Canadair CL-44s from 1962. In October 1962 scheduled flights were resumed, when Slick Airways was granted permanent permission to operate its transcontinental Route 101. The trunk routing went from either San Francisco or Burbank to Dallas, St. Louis, Chicago, Indianapolis and New York City and was operated using the Lockheed Super Constellation. The DC-6 was used on side routes. Additionally, Slick Airways operated on so called Quicktrans domestic routes on behalf of the United States Navy. In 1965, the L-1049 was withdrawn from the Route 101 in favor of the CL-44. On 27 August 1965, scheduled flights once again were discontinued because of the poor financial situation of Slick Airways, and military charter flights were terminated by the end of the year. On 1 July 1966, Slick Airways was shut down, and the assets were acquired by Airlift International. (27AUG2022) https://en.wikipedia.org/wiki/Slick_Airways

SPAM: Spam (stylized as SPAM) is a brand of canned cooked pork made by Hormel Foods Corporation. It was introduced by Hormel in 1937 and gained popularity worldwide after its use during World War II.By 2003, Spam was sold in 41 countries on six continents and trademarked in over 100 countries. Spam's basic ingredients are primarily pork shoulder and ham,with salt, water, modified potato starch (as a binder), sugar, and sodium nitrite (as a preservative). Natural gelatin is formed during cooking in its tins on the production line.[4] Concerns about Spam's nutritional attributes have been raised, in large part due to its high content of fat, sodium, and preservatives. (27AUG2022) https://en.wikipedia.org/wiki/Spam_(food)

tarmac: tarmac, incorrectly used as a term for any paved surface of an airport, for example

Tatoo: "Tattoo" is a bugle call played in the evening in the British Army and the United States Army. The original concept of this call was played on the snare drum and was known as "tap-too", with the same rule applying. Later on, the name was applied to more elaborate military performances, known as military tattoos. The etymology of the military tattoo is from Dutch "tap toe", unrelated to the Tahitian origin of an ink tattoo. (27AUG2022) https://en.wikipedia.org/wiki/Tattoo_(bugle_call)

The Hump: he Hump was the name given by Allied pilots in the Second World War to the eastern end of the Himalayan Mountains over which they flew military transport aircraft from India to China to resupply the Chinese war effort of Chiang Kai-shek and the units of the United States Army Air Forces (USAAF) based in China. Creating an airlift presented the USAAF a considerable challenge in 1942: it had no units trained or equipped for moving cargo, and no airfields existed in the China Burma India Theater (CBI) for basing the large number of transports that would be needed. Flying over the Himalayas was extremely dangerous and made more difficult by a lack of reliable charts, an absence of radio navigation aids, and a dearth of information about the weather.The task was initially given to the USAAF's Tenth Air Force, and then to its Air Transport Command (ATC). Because the USAAF had no previous airlift experience as a basis for planning, it assigned commanders who had been key figures in founding the ATC in 1941–1942 to build and direct the operation, which included former civilians with extensive executive experience operating civil air carriers.Originally referred to as the "India–China Ferry",[4] the successive organizations responsible for carrying out the airlift were the Assam–Burma–China Command[a] (April–July 1942) and the India-China Ferry Command (July–December 1942) of the Tenth Air Force; and the Air Transport Command's India-China Wing (December 1942 – June 1944) and India-China Division (July 1944 – November 1945). The operation began in April 1942, after the Japanese blocked the Burma Road, and continued daily to August 1945, when the effort began to scale down. It procured most of its officers, men, and equipment from the USAAF, augmented by British, British-Indian Army, Commonwealth forces, Burmese labor gangs and an air transport section of the Chinese National Aviation Corporation (CNAC). Final operations were flown in November 1945 to return personnel from China. The India–China airlift delivered approximately 650,000 tons of materiel to China at great cost in men and aircraft during its 42-month history.For its efforts and sacrifices, the India–China Wing of the ATC was awarded the Presidential Unit Citation on 29 January 1944 at the personal direction of President Franklin D. Roosevelt,the first such award made to a non-combat organization.(27AUG2022) https://en.wikipedia.org/wiki/The_Hump

UFO: Unidentifed Flying Object

Unlimiteds: With the exception of very few "scratch-built" aircraft, the Unlimited Class has generally been populated by stock or modified WWII fighters with the P-51 Mustangs, F-8F Bearcats and Hawker Sea Furies being flown most often. The Unlimited Class flies in speeds exceeding 500 mph. (27AUG2022) https://airrace.org/sights-and-sounds/racing-class-info/unlimited-class-gallery/

V' formation: A V formation is the symmetric V-shaped flight formation of flights of geese, swans, ducks, and other migratory birds, improving their energy efficiency. Usually, large birds fly in this formation since smaller birds create more complex wind currents that are hard for the birds in the back to take advantage of.V formations also improve the fuel efficiency of aircraft and are used on military flight missions. (27AUG2022) https://en.wikipedia.org/wiki/V_formation

WX: Weather, in Morse code shorthand. (27AUG2022) https://en.wikipedia.org/wiki/WX

www.ingramcontent.com/pod-product-compliance
Lightning Source LLC
LaVergne TN
LVHW070107250926
848802LV00016B/5
9781630100308